course back up for those questioned for assertiveness & leadership. Sponsored by companies/universities

BALANCING THE [illegible]

Gender Issues in the Building Professions

: thoughts

↓

14-16 –
16-18 – 2.

Questionnaire to school children ask Q's about their aspirations and about their parents careers

↓

See if connection to wealth/job influences.

Possibly ask for names to further follow up of those interested in construction related careers. Follow through to see how the females do over time.

Other titles from IES:

Women into Management
Wendy Hirsh, Charles Jackson
Report P158. 1991. 1-85184-110-5

Beyond the Career Break
Wendy Hirsh, Sue Hayday, Jill Yeates, Claire Callender
Report 223. 1992. 1-85184-146-6

Family Friendly Working: Hope or Hype?
Jim Hillage, Clare Simkin
Report 224. 1992. 1-85184-147-4

A full catalogue of publications is available from IES.

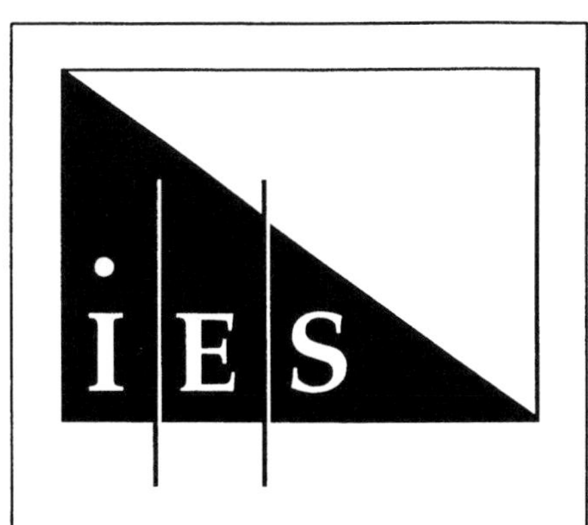

BALANCING THE BUILDING TEAM: GENDER ISSUES IN THE BUILDING PROFESSIONS

Gill Court, Janet Moralee

THE INSTITUTE FOR EMPLOYMENT STUDIES

Report 284

Published by:

THE INSTITUTE FOR EMPLOYMENT STUDIES
Mantell Building
University of Sussex
Brighton BN1 9RF
UK

Tel. + 44 (0) 1273 686751
Fax + 44 (0) 1273 690430

A catalogue record for this publication is available from the British Library

ISBN 1-85184-210-1

Printed in Great Britain by Microgen UK Ltd

The Institute for Employment Studies

The Institute for Employment Studies is an independent, international centre of research and consultancy in human resource issues. It has close working contacts with employers in the manufacturing, service and public sectors, government departments, agencies, professional and employee bodies, and foundations. Since it was established 25 years ago the Institute has been a focus of knowledge and practical experience in employment and training policy, the operation of labour markets and human resource planning and development. IES is a not-for-profit organisation which has a multidisciplinary staff of over 50. IES expertise is available to all organisations through research, consultancy, training and publications.

IES aims to help bring about sustainable improvements in employment policy and human resource management. IES achieves this by increasing the understanding and improving the practice of key decision makers in policy bodies and employing organisations.

Formerly titled the Institute of Manpower Studies (IMS), the Institute changed its name to the *Institute for Employment Studies* (IES) in Autumn 1994, this name better reflecting the full range of the Institute's activities and involvement.

Acknowledgements

The research on which this report is based is very much the product of a joint effort between the IES researchers, the project Steering Group, and the women and men who agreed to participate in the study. We would like to thank all those who gave their time and expertise to the project. The support and organisational skills of Lucky Lowe, Louisa Sheppard (CITB) and Frank Lettin (CIOB) were of immense value. Ian Mackay (CITB) and Sally Kirk-Walker (Institute of Advanced Architectural Studies, University of York) generously provided help with the literature review, while Katherine Bowyer of the CIOB Information Resource Centre tracked down some of the more difficult to find documents we were looking for.

Within IES, we would like to thank Emma Hart, who took care of the administration of the project and helped to prepare this report, Andy Davidson and Louise Paul. We are also grateful to the IES Survey Unit, led by Monica Haynes, for their work on the postal survey. Finally, Sally Dench provided valuable insights into the issues we were addressing.

Contents

Action Plan

To achieve the potential benefit of a balanced workforce the building industry needs to take the following actions:

- **Promote public awareness of the role of women in building**

The building industry is not strongly associated with the employment of women. Raising the profile of women in the industry will help young people realise that women do have a role in building and that it can offer rewarding careers for women and men.

- **Support educational initiatives in building**

Building needs to be seen as an activity of universal relevance. Educational initiatives help to raise the general awareness of the role of construction in society, making it relevant to all young people and thereby encouraging both girls and boys to consider a career in the industry.

- **Promote general school and college liaison activities**

If young women are to consider a career in the building professions they need to know more about the industry and to see women working within it. Involving women in school and college liaison activities will help overcome the perception that professional careers in building are not suitable for women.

- **Offer work placements and work shadowing for girls and young women**

Few girls and young women know what it would be like to work as a building professional. Work placements and work shadowing provide young people with an insight into the variety of careers the building industry has to offer. This kind of first hand experience can overcome the general lack of knowledge of career opportunities in building and counter the industry's negative image.

- **Offer work placements and work shadowing for teachers and careers advisers**

Teachers and careers advisers are a major source of information on potential careers. Providing these advisers with direct experience of the building industry will help raise awareness of the range of careers available in building.

- **Improve parents', teachers' and careers advisers' knowledge of careers in building**

There is a general lack of awareness about building careers and what they can offer to both young women and young men. Accessible, widely distributed, information on the careers available in the industry would help overcome one of the major barriers to increased participation by women.

- **Make the business case for equal opportunities in the building industry**

If barriers to the full participation of women in the building industry are to be dismantled, employers need to be convinced of the arguments for taking action. Establishing the business case for equal opportunities will encourage more employers to act to change the current situation.

- **Ensure the principle of equal opportunity is followed in recruitment activities**

If more women are to enter careers in the building professions, employers need to be aware of how their recruitment literature and recruitment processes affect potential female employees. Following equal opportunities practices in recruitment will encourage women to apply for positions and ensure that their applications are considered fairly.

- **Help older women enter a professional career in building**

Existing education and training opportunities leading to a professional role in building are aimed mainly at younger men and women. Encouraging the industry to take on mature women trainees and newly qualified workers will allow employers to tap into a rapidly growing part of the working population.

- **Support women entering and pursuing careers in building**

There are currently few women in the building professions. Providing support for those entering and working in the industry will help counter some of the isolation which women may experience. It will also allow women to share the positive experiences of colleagues and to deal better with difficult situations.

- **Provide women with the necessary skills to deal with difficult people**

Women and men in building need to be able to deal with some of the aggressive and confrontational attitudes which pervade parts of the industry. Inter-personal skills training will equip people with the necessary skills to deal with difficult individuals and thereby reduce the impact of these encounters.

- **Counter the negative image of women-only events**

Women in building view women-only events in a negative light. The more positive aspects of these activities need to be highlighted if their benefits are to be fully realised.

- **Ensure equal opportunities are taken seriously by building employers**

There is a mixed attitude toward equal opportunities in the building industry. If women are to be encouraged to enter and remain in the industry this attitude has to change. Of particular importance are commitment from the top of the organisation and evidence of the implementation and monitoring of an equal opportunities strategy.

- **Enable women to combine work and family life**

Currently the building industry makes it difficult for women to combine work and family life. If the industry is serious about increasing the representation of women at all levels, it must promote the provision of affordable and accessible childcare and a range of 'family friendly' working arrangements.

About the report

This report presents the findings of a study commissioned by the Chartered Institute of Building (CIOB) and the Department of the Environment on how best to improve the representation of women in professional, managerial and technical occupations in the building industry. A key aim of the research was to develop a series of recommendations to encourage more women to enter professional careers in building and subsequently to remain within the industry. The research was conducted between July 1994 and February 1995.

1. Introduction

This report presents the findings of a study commissioned by the Chartered Institute of Building (CIOB), and supported by the Department of the Environment (DoE), on how best to improve the representation of women in professional, managerial and technical occupations in the building industry. A key aim of the research was to develop a series of recommendations to encourage more women to enter building careers in the first place and subsequently to remain within the industry.

1.1 Background

1.1.1 The under-representation of women in building

The under-representation of women in the building industry began to be highlighted as an issue in the 1980s. This period has seen a number of initiatives to raise awareness of the issue and act to improve the current situation. These include UCATT's (Union of Construction and Allied Trade Technicians) Women Working in Construction Committee and *Blueprint for Equality*; Women and Manual Trades (WAMT); Women's Education in Building; Women as Role Models (WARM); the former Women in Construction Advisory Group (WICAG); the Women in Construction Alliance; the Equal Opportunities Task Force of the Latham Review; and the CIOB's Women in Building Committee (WIBC).[1]

The WIBC was formed in 1989 as a forum for discussion about the barriers to women's entry into building careers. WIBC's role was later extended and one of its first actions was to undertake an assessment of women's contribution to the Institute. The aim of this research was to better understand the experiences of women members of the CIOB and to raise awareness of women's expertise amongst the general membership.

The results of this initial study were published as *His House or Our House?* in 1993 (Lowe and Byrne, 1993). The report documents trends in women's membership of CIOB and points

---

[1] see Dainty, 1993; Women's Education in Building (n.d.); EOR, 1990; Lowe and Byrne, 1993; Latham, 1994 for details.

out that female graduates are the fastest growing membership category, fuelling an increase in the number of women members which is faster than that for men. This is an encouraging trend but further investigation of women's experience in the building industry revealed another side to the story. Almost a half of those surveyed thought that they had suffered discrimination at work, whilst others described a series of incidents which clearly indicated that women were treated far from equally (Lowe and Byrne, 1993).

In order to investigate further these issues and to generate recommendations for improving the current situation, the CIOB obtained DoE support and commissioned IES to undertake the study which is reported here.

1.1.2 Why is it an issue?

There are two key reasons why it is worth investigating women's under-representation in the building industry. The first concerns the changing pattern of skill demand and the need to attract a high quality work force. The second relates to the prospect of generally improving the way the industry operates and its associated working environment.

Work force quality

It is only in the past two decades or so that graduates have been able to circumvent the traditional route into construction management involving an apprenticeship in trades followed by moves up to site supervisor, site management and beyond (Lowe and Byrne, 1993; IPRA, 1991). These routes are now being replaced by entry through more formal professional training, as indicated by the increase in the number of students on construction degree programmes and the move to make graduate status a requirement for membership of relevant professional organisations (IPRA, 1991; CITB, 1994a).

Underlying this move to increase the proportion of graduates in the industry is a recognition of the value they add to work in terms of productivity and efficiency. The increasing technical complexity of construction projects alongside changes in how organisations co-ordinate their activities mean that managers and professionals require both general skills and knowledge, including team work and communication skills, and the ability to integrate a range of technical and managerial skills (IPRA, 1991). This requires a high level of training and ability and the industry is currently only tapping a small proportion of individuals with the potential to develop the skills it is seeking. Women now account for almost a half of the total population of new graduates and they have already successfully entered a range of previously male dominated professions: these professions' gain is the building industry's loss.

There are also indications that women's skills in some of these areas are better developed, in particular the core skills of presentation and communication (Chevin, 1995).

In addition, as women increase their representation in a range of activities, including the law, accountancy, and management, they will form a higher proportion of those with whom building industry managers have to negotiate (Chevin, 1995). This client-driven aspect of equal opportunities has already provided the impetus for improving prospects for women in other male dominated environments, in particular engineering (Opportunity 2000, 1994; Wilkinson, 1992a).

Culture and working environment

There are two key issues here. First, that the current working environment can be as problematic in attracting high calibre men to, and keeping them within, the building industry as it is for women (see Gale, 1991a; 1992).[1] This is the case in both the craft and professional areas of the industry, but with their higher level of qualifications it is likely that young professionals will have a broader range of options in the labour market.

Second, the Latham Review highlighted the cost of the adversarial attitudes which dominate the industry (Latham, 1993; 1994). These are paralleled by a general culture of conflict within the industry (Gale, 1992; Sommerville and Stocks, 1993) in which good relationships with clients and co-workers are undervalued. In the context of the changes proposed by the Latham Review, a move toward less conflictual forms of interaction is possible. Employers in the industry have already highlighted women's strengths in this area, noting that they get on better with clients, create a better working environment and are able to change the attitudes of men on-site (Chevin, 1995).

In sum, not only do women account for a growing proportion of the highly qualified workforce but they also demonstrate to a greater extent the kinds of skills increasingly in demand.

In recognition of the need to widen the recruitment base in construction, one of the recommendations of the Latham Review was the establishment of an Equal Opportunities Task Force. Its remit is to develop an action-plan for improving the representation of women in the industry (New Builder, 1994). This is due to report later this year and is a clear indication of the importance accorded the issues addressed in this report.

---

1 As the author of one recent article put it: 'Good people are leaving the trade every week to work as market traders or taxi drivers or they are going abroad where conditions are better' (Barr, 1994).

1.2 Objectives

The research aimed to examine women's experience of working in the building industry and to develop suggestions for specific measures which could be introduced to improve the representation, retention and progress of women in building. A key dimension was how best to create a working environment in which all groups of workers (both men and women) are able to work together to achieve their full potential. In addition, the study looked at:

- existing research on why women are poorly represented in building and other male dominated professions
- the factors influencing women's entry into building
- the reasons women give for leaving the industry
- their views on working in building and experience of building related courses.

The focus of the analysis was people in professional and technical roles. The CIOB takes the view that there are no professional and technical roles in the industry which cannot be undertaken by women. Women's marked under-representation at this level (they account for just two per cent of CIOB members) is therefore a particular cause for concern.

1.3 Approach

Our approach to the study was to canvass the views of women with experience in the building industry. We were specifically concerned to base our recommendations on what they thought would be the most appropriate actions to take in order to increase the number of women entering and remaining in building. As a result, we adopted a three part research strategy (see Appendix 1 for details):

1. a literature review to bring together existing research on women in construction and similar industries. The main aim here was to understand the reasons for women not considering careers in such industries in the first place and to summarise existing knowledge on why they do not remain within them or fail to progress at the same rate as their male counterparts.
2. a postal survey of women members of CIOB in order to establish their views on working in building and what could be done to improve the current situation.
3. a series of discussions with respondents to the survey, employers, careers advisers and educationalists, and key industry figures. The aim of this part of the study was to gain a more in-depth understanding of the issues emerging

from the postal survey results and to gauge industry reaction to a range of proposals.

1.4 Structure of the report

This report is divided into seven main sections. The first two summarise existing research on women in male dominated occupations, including the construction industry, and include:

- an overview of the labour market context, including summary statistics on women's employment in the construction industry
- a discussion of the barriers to increasing the representation of women in building, including a review of research on why women do not choose to enter the construction industry, and the problems they encounter if they do embark on such careers.

The next four chapters detail the findings of the postal survey and discussion groups conducted as part of this study. They each cover a different topic, as follows:

- details of the respondents to the survey, including their personal and educational qualifications, and current employment
- the factors influencing career decisions (entering and leaving the industry)
- the experience of women on building related courses
- women's views on the industry and their assessment of actions to improve the current situation.

The final chapter draws on the results of the study, including discussions with all the groups involved in it, to develop a series of recommendations for change.

2. The Labour Market Context

This chapter provides the background for the rest of the study by briefly summarising the main features of the labour market for women. The primary focus is their representation in the labour market for technologists, engineers and scientists. The chapter is divided into three main sections. The first looks at some of the trends in the labour market, and women's participation within it, over the past two decades. The second outlines the findings of recent research on women in technology and science. The final part of the chapter looks at the situation in the construction industry.

2.1 Women at work

There is an immense amount of literature on women's changing experience of the labour market. We could not possibly attempt to summarise it fully in a single chapter and have therefore chosen to highlight some key issues of particular relevance to this study. They include:

- changes in economic activity
- the distribution of women's employment by occupation
- rising educational attainment.

Key figures on each of these issues are highlighted in the accompanying boxes.

2.1.1 Economic activity

The key issue here is that the vast majority of women of working age are in the labour force (over 70 per cent), a pattern which is projected to continue well into the next century. Women's rising rates of economic activity[1] contrast with the pattern among men, a declining proportion of whom are in the labour force (see Box 2.1). There are two main implications of this trend:

---

[1] The economic activity rate is the proportion (per cent) of all people of working age who are either in employment or who are unemployed/looking for work. In this report the term 'economic activity' is used interchangeably with that of 'labour force participation'.

Box 2.1 Economic activity: key trends

- The economic activity rate of women increased from 57 per cent in 1971 to 71 per cent in 1994, and is projected to rise to 75 per cent in 2006.
- Equivalent figures for men show a decline from 91 to 84 per cent between 1971 and 1994, and to 82 per cent by 2006.
- Women's share of all employment has therefore increased: they are projected to account for 46 per cent of the labour force in 2006, up from 37 per cent in 1971.
- The main increase has been in the proportion of mothers who work: in 1973, less than half (47 per cent) of women with dependent children were in paid work, a figure which had increased to 59 per cent by 1992.
- The most marked change has occurred among women with children under five, more than two-fifths of whom are now in the labour force compared to only a quarter in 1973.
- These patterns are a result of a marked increase in the rates at which mothers have been returning to paid work within a few months of having a baby. By the late 1980s, nearly half of mothers who were in work when they became pregnant were back at work within nine months of having their babies. A decade earlier, in the late 1970s, only a quarter had returned to work within nine months of giving birth.
- The proportion of women with no dependent children who are working has risen much more slowly, from 69 per cent in 1973 to 72 per cent in 1992.
- These changes mean that one in every six employed people is a woman with a child under the age of 16.

Sources: Court, 1995; McRae and Daniel, 1991; Sly, 1994

1. Women's share of total employment is rising and they account for a particularly high proportion of new entrants to the labour force. Recent projections suggest that by the year 2006, over 45 per cent of the labour force will be female. In addition, women are expected to account for 80 per cent of the increase in the labour force between 1994 and 2006 (Ellison, 1994). The obvious implication here is that employers must look to both women and men if they are to make full use of the potential in the labour market.
2. A rising proportion of people in the labour market are mothers, in particular mothers of young children. The increase in women's labour market participation highlighted above has been driven almost entirely by a transformation in the working patterns of women with children (see Box 2.1 for details). Given that women remain largely responsible for raising children, employers who make it easier for them to combine family life with working will be in a better position to attract and retain women employees.

2.1.2 Occupational distribution of women's employment

Women's labour market participation patterns are increasingly similar to those of men. The two groups remain, however, very different in terms of the types of jobs they do. Women are over-represented in lower level non-manual occupations such as clerical and secretarial work, personal services and sales occupations. They are under-represented in managerial jobs, some professions (in particular, engineering and science) and in manual occupations (Wilson, 1994).

Over the past two decades, however, one of the main changes in women's employment has been their entry into professional and managerial positions. In 1971 these occupational categories accounted for just 12 per cent of women, a figure which had increased to 20 per cent by 1993. As a result, women's share of total employment has risen in each of the major managerial and professional occupations (Table 2.1). Within these occupational categories, the change has been particularly marked among 'corporate managers and administrators', and the 'other professional' and 'other associate professional' occupational

Table 2.1 Occupational profile of women and men

	1993		Ratio women: men[1]	Women's share of total employment		
	Distribution o women (%)	Distribution of men (%)	1993	1971	1981	1993
Managers & administrators	12.4	20.2	0.61	20.8	23.7	34.3
Professionals	8.1	10.4	0.80	32.1	34.3	40.7
Science & engineering professionals	*0.6*	*3.7*	*0.15*	*2.5*	*6.2*	*11.7*
Associate professional/technical	10.4	8.3	1.25	40.6	45.5	51.5
Science & engineering associate prof./technical	*1.1*	*3.5*	*0.32*	*10.8*	*15.1*	*21.4*
Clerical & secretarial	27.4	6.7	4.12	66.4	72.0	77.8
Craft & related	3.5	22.4	0.16	12.7	11.0	11.7
Personal & protective service	13.7	5.5	2.49	62.0	62.9	68.0
Sales	10.4	4.6	2.24	53.5	60.7	65.6
Other	14.2	22.2	0.64	32.1	36.4	35.3
Total	100.0	100.0	1.00	36.5	40.6	46.0

Source: Lindley and Wilson, 1983; 1984

[1] This is the ratio of column 2 (distribution of women's employment) to column 3 (distribution of men's employment) and provides an indication of the extent to which women are over- or under-represented in the occupation relative to men (a figure of less than 1 implies under-representation and one of greater than 1 implies over-representation).

Box 2.2 Women in the professions

- **medicine** – women now account for almost a half of medical students, up from just over a fifth in the 1960s
- **law** – about a half of new entrants are women, up from less than a fifth in the mid-1970s
- **banking** – in 1970 women accounted for just two per cent of successful finalists in the Institute of Banking exams, a proportion which had risen to 27 per cent by the late 1980s.
- **accountancy** – in the early 1970s only three per cent of new chartered accountants were women. By the early 1990s, this figure had increased to almost a half (48 per cent).

Source: see Court, 1995.

groups. These include the law, accountancy and banking.[1]

Underlying these trends are substantial changes in women's representation among new entrants to a range of previously male dominated professions (see Box 2.2). The main implication of these patterns is to demonstrate that the gender mix within occupations is not necessarily fixed — large scale transformations are possible and have occurred in the relatively recent past. As discussed in 2.2, however, the rate of change varies markedly, with some areas of activity remaining heavily male dominated.

2.1.3 Educational attainment

One of the reasons for women's increased labour market participation and their entry into higher level occupations is rising levels of educational attainment (see Box 2.3). There has been a general increase in the qualification levels of school leavers in recent years, but young women now surpass men in terms of the proportion leaving with 2 or more 'A' levels (26 per cent did so in 1992 compared with 23 per cent of men). The success of women at gaining qualifications is likely to continue in the near future, if only because of their achievements earlier in the secondary school system. In 1993, 46 per cent of girls gained five or more GCSEs at grades A-C compared to 37 per cent of boys in the same age group (DfE Statistical Bulletin, 1994a).

The main area in which women's attainment continues to lag behind that of men is in vocational qualifications. While both groups are now almost equal in terms of the proportion on vocational courses, women are poorly represented among those

---

[1] Women remain, however, under-represented in the top levels of all occupations, even those which are heavily female dominated. For example, they account for almost three-quarters of full-time teachers but comprise just 21 per cent of secondary school head teachers (Court, 1995).

Box 2.3 Women's educational attainment

- In the mid-1960s just nine-per cent of female school leavers left with 2 or more 'A' level qualifications, a figure which had increased to 26 per cent by the early 1990s (among male leavers the proportion rose from 13 to 23 per cent).
- The GCSE attainment of young women (aged 15) is now well above that of young men (in 1993, 46 per cent gained five or more GCSE at grades A to C compared with 37 per cent of young men).
- In 1992, women accounted for 46 per cent of full-time first degree enrolment and 38 per cent of full-time postgraduate students.
- In 1992, 45 per cent of all university first degrees were awarded to women, up from 27 per cent in the mid-1960s.
- Over the same time period, women's share of university higher degrees also increased (from nine per cent to 36 per cent).
- Women's attainment of vocational qualifications at NVQ level 3 is substantially lower than that of men — in 1993, more than 20 per cent of young men had such qualifications at this level compared to a little over 10 per cent of women.

Source: Gibbins, 1994; NACETT, 1994; DfE Statistical Bulletin, 1994a; Court and Meager, 1994

who gain NVQ Level 3 qualifications and above. This is related to the kinds of qualifications they obtain. Young women are more likely to gain RSA qualifications (which are mainly offered in clerical and related subjects), the majority of which are gained at NVQ level 1 or its equivalent. Young men, on the other hand, are more likely to gain a BTEC or City and Guilds qualification (mainly offered in craft and technical subjects), the majority of which are gained at NVQ level 2 or above (Gibbins, 1994).

Changes in women's attainment of secondary school qualifications have been mirrored within the higher education system. Women now account for almost a half of all higher education enrolment, up from less than a quarter in the early 1960s (Court and Meager, 1994. They also account for an increasing proportion of postgraduate students (38 per cent in 1992).

2.2 Women in technology and science

We noted above that women's entry into some higher level occupations had proceeded at a faster pace than in others. This section looks in more detail at engineering, technology and scientific occupations, which remain highly male dominated.

There are two dimensions to the employment of women in technology and science in the UK. First, women remain a small proportion of the number of people employed in these jobs. Second, even when they are present, women are over-represented at the lower end of the relevant career hierarchies and under-represented in more senior positions (Committee on Women in Science, Engineering and Technology, 1994).

2.2.1 Women's employment in technology, engineering and science

Overall, about 65,000 women work in engineering and science professions, accounting for almost 12 per cent of employment in this occupational category (this compares with women's share of 46 per cent of employment in all occupations). Almost double this number work in the engineering and science 'associate professional' occupational category (123,000), a group which includes laboratory assistants and other technicians. In these occupations women account for 21 per cent of total employment, a figure which is considerably higher than in similar professional activities but still lower than in the workforce as a whole. In total, just two per cent of all women in employment work in these two occupational categories, a figure which rises to seven per cent for men (see Table 2.1 above).

Looking in more detail at this set of occupations[1], it is clear that women are concentrated in the biological and life sciences with fewer working in the physical sciences and engineering. They outnumber men in only one occupation: that of laboratory technician (Committee on Women in Science, Engineering and Technology, 1994, Figure 2.4).

The situation has, however, improved since 1981 when SET occupations employed just 101,000 women. This means that between 1981 and 1993 the number of women working in engineering and science increased by 86 per cent. Over the same period total female employment expanded by 15 per cent. As a result, the overall proportion of all women who work in SET occupations has risen from one to two per cent (Table 2.1).

These data show that an increasing, though still small, number of women are entering SET occupations. This is reflected in the age profile of men and women in employment who have qualifications in SET subjects. Those aged 25 to 34 dominate the female age profile, while the male profile shows a much broader distribution over the 25 to 49 age range (Committee on Women in Science, Engineering and Technology, 1994, Figure 9).

A second reason for the relatively youthful age profile of working women with SET qualifications may be due to their higher propensity to leave after a few years in employment. The number of women peaks in the 25 to 29 age range. Breaking the data down by part-time and full-time work shows that the

---

[1] In the remainder of this report technology, engineering and science professional and associate professional occupational categories are referred to collectively as SET occupations. This is the abbreviation used in the Committee on Women in Science, Engineering and Technology (1994) report, from which much of the following information is derived.

number working part-time reaches a maximum at age 35-39. This increase in part-time work is not, however, sufficient to compensate for the decline in full-time working which takes place from age 30 onwards, a pattern which suggests that a number of women qualified in SET disciplines are lost to the workforce in their thirties (Committee on Women in Science, Engineering and Technology, 1994, p.16). This coincides with the time many women will be choosing to have children.

International comparisons

Comparing the UK with other countries shows that while women are not generally equally represented in science and technology there is variation in their share of these activities (National Science Board, 1993). In the former USSR and Eastern Europe women accounted for a higher proportion of engineers and scientists than is the case generally in Western Europe, while in the US, France and Egypt the proportion of women in engineering is higher than in the UK (Chivers, 1988; Lowe and Byrne, 1993; Wilkinson, 1992a; Rees, 1994; National Science Board, 1993). This suggests that their low representation in the UK can not be fully explained by natural limitations or inclinations.

2.2.2 Women's position in the occupational hierarchy

A number of recent publications have shown that the representation of women in many industries declines as the seniority of the post increases (Committee on Women in Science, Engineering and Technology, 1994, p.16; Personnel Management, 1994; Hansard Society Commission, 1990; Davis, 1994). For example, in 1992 women accounted for 21 per cent of scientists in the Civil Service. Their representation, however, varied substantially by grade, from 35 per cent at Assistant Scientific Officer level to just nine per cent in Senior Scientific Officer posts. A similar pattern was evident among Professional and Technology Officers, just three per cent of whom were women. These were concentrated in the Technician Grades and account for less than two per cent of those at Scientific Officer level and above. The one encouraging piece of data is that almost 19 per cent of trainees are women, which suggest that the situation may change in the future (Committee on Women in Science, Engineering and Technology, 1994, p.19).

This picture is repeated in a number of other areas of activity, including:

- higher education, where even in biology (a science with better than average representation of women) less than five per cent of professors are women
- learned societies and professional organisations (women comprise less than five per cent of fellows in most of the

major SET professional institutions, the main exception being the Institute of Biology)

- industry (only eight per cent of employees in the top four levels of industrial organisations surveyed as part of the *Rising Tide* report were women) (Committee on Women in Science, Engineering and Technology, 1994).

2.3 Women in the construction industry

The construction industry is a particular example of the full range of technological and scientific activities and many of the features outlined above are represented within it. If anything, these features are especially prevalent in construction. This section summarises data on women in the construction industry, with particular reference to building. The discussion is divided into two main parts:

- data on employment
- representation in professional organisations.

2.3.1 Women's employment in construction

In 1994 there were about 1.77 million people working in the construction industry, just 174,000 (less than 10 per cent) of whom were women. This makes the industry the most male dominated of all the major industry groups (Table 2.2).

Women's employment in the construction industry did however increase over the decade to 1994 (by 14 per cent), whereas that for men declined (by seven per cent). As a result, women's share of total employment rose slightly.

Table 2.2 Employment (employees and self-employed) by industry and gender: Spring 1994 (thousands unless specified)

	Women	Men	Total	% women
Agriculture, forestry & fishing	109	355	464	23.5
Energy & water supply	64	272	336	19.0
Manufacturing	1316	3388	4704	28.0
Construction	174	1592	1766	9.9
Distribution, hotels & catering	2571	2358	4929	52.2
Transport & communication	322	1202	1534	21.0
Banking, finance, insurance etc.	1590	1776	3366	47.2
Public admin, education & health	4069	1879	5948	68.4
Other services	783	621	1404	55.8
Total	11226	13716	24942	45.0

Source: Employment Gazette, December 1994, p.LFS3

Table 2.3 Employment in construction by occupation and gender: 1991 (per cent)

	Distribution of men	Distribution of women	Women's share of total
Managers & administrators	9.3	15.4	14.0
Professional occupations	2.9	1.1	3.4
Associate professional & technical	2.9	2.6	8.0
Clerical & secretarial	1.1	63.0	84.5
Craft & related	61.2	4.8	0.8
Personal & protective service	0.2	0.6	22.3
Sales	0.8	4.7	36.7
Plant & machine operatives	9.9	1.8	1.8
Other occupations	11.0	5.6	4.8
Total	**100.0**	**100.0**	**9.0**

Source: 1991 Census: Great Britain Economic Activity, Table 12

These data, however, only tell part of the story. Of greater interest for the purposes of this study is the kind of work women in the industry do. Table 2.3 shows that the vast majority of women in construction work in a clerical and secretarial capacity (63 per cent), while for men the 'craft and related' occupational category dominates employment. Looking at women's share of total employment in each occupation, women are over-represented in the 'clerical and secretarial', 'sales', and 'personal and protective services' categories and under-represented elsewhere. They are particularly poorly represented in manual occupations (especially the 'craft and related', and 'plant and machine operatives' categories) and professional occupations.

Table 2.3 also indicates that women are much better represented in the 'associate professional and technical' occupational category than in the professions themselves (accounting for eight per cent of employment in the former and just three per cent in the latter).

More detailed investigation of the three top categories reveals further differences between women and men (Figure 2.1). First, in the 'managerial and administrative' category, women's share of total employment is highest among 'specialist managers' (a category which includes personnel and public relations managers), and lowest among 'production managers'. One of the findings of research on women in higher level occupations in recent years is that they tend to be channelled into specialist posts and not into mainstream management (see Chapter 3 for a more detailed discussion). These data indicate that a similar pattern has emerged in the construction industry.

Second, within professional occupations, women are less well represented in engineering and technology occupations than their male counterparts (74 per cent of men in construction

Figure 2.1 Women's share of total employment in selected construction industry occupational categories: 1991

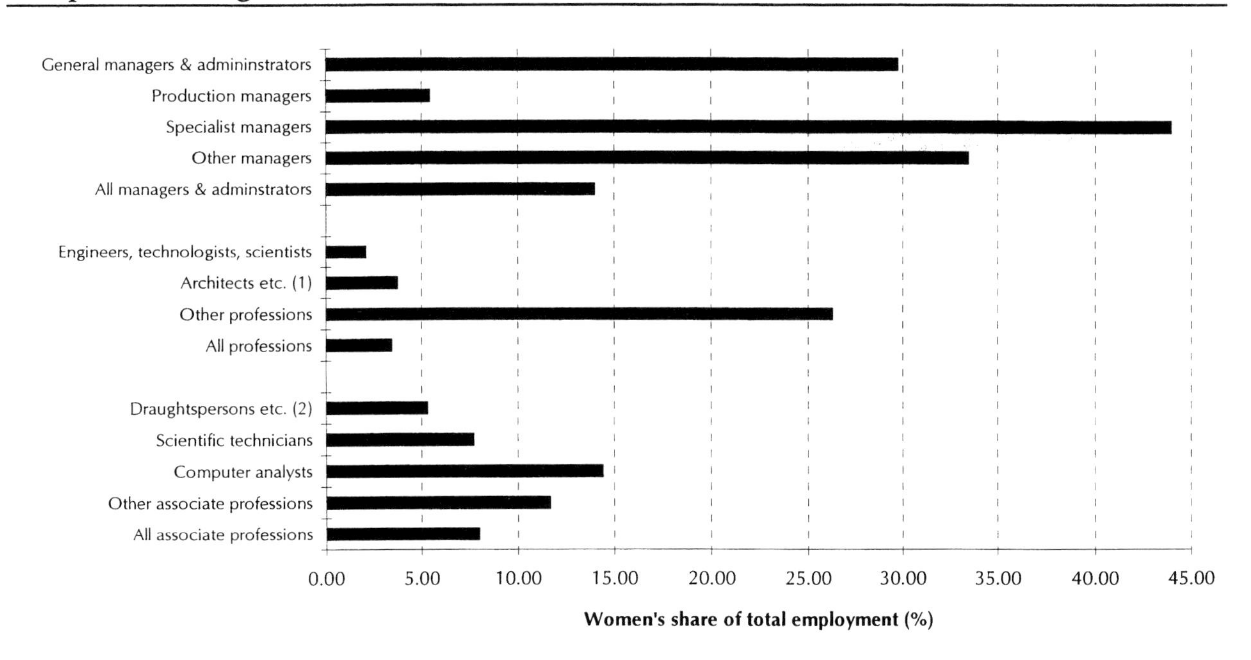

Source: 1991 Census, Economic Activity: Great Britain, Table 4

1. This category includes architects, town planners and building, land, mining and 'general practice' surveyors
2. This category includes draughtspersons, building inspectors, quantity surveyors, and marine, insurance and other surveyors.

industry professions are engineers and technologists compared to 43 per cent of women). As a result, women's share of employment in the professional engineering, technologist and scientist category is just two per cent (Figure 2.1). On the other hand, they are better represented in the 'other professions' category, which includes business and financial professionals and legal, health and teaching professionals.

In the 'associate professional' occupations, women's share of total employment is also highest in non-technical activities such as the law, business and finance, and health (categorised in Figure 2.1 as 'other associate professions'). They are less well represented in the 'draughtspersons, building inspectors, quantity and other surveyors' category.

This is consistent with data on the building industry which show that women are concentrated in somewhat different parts of the industry to men, being better represented in some areas of activity than others (research and education, environmental

Figure 2.2 Primary activity of CIOB members: women as a per cent of total membership

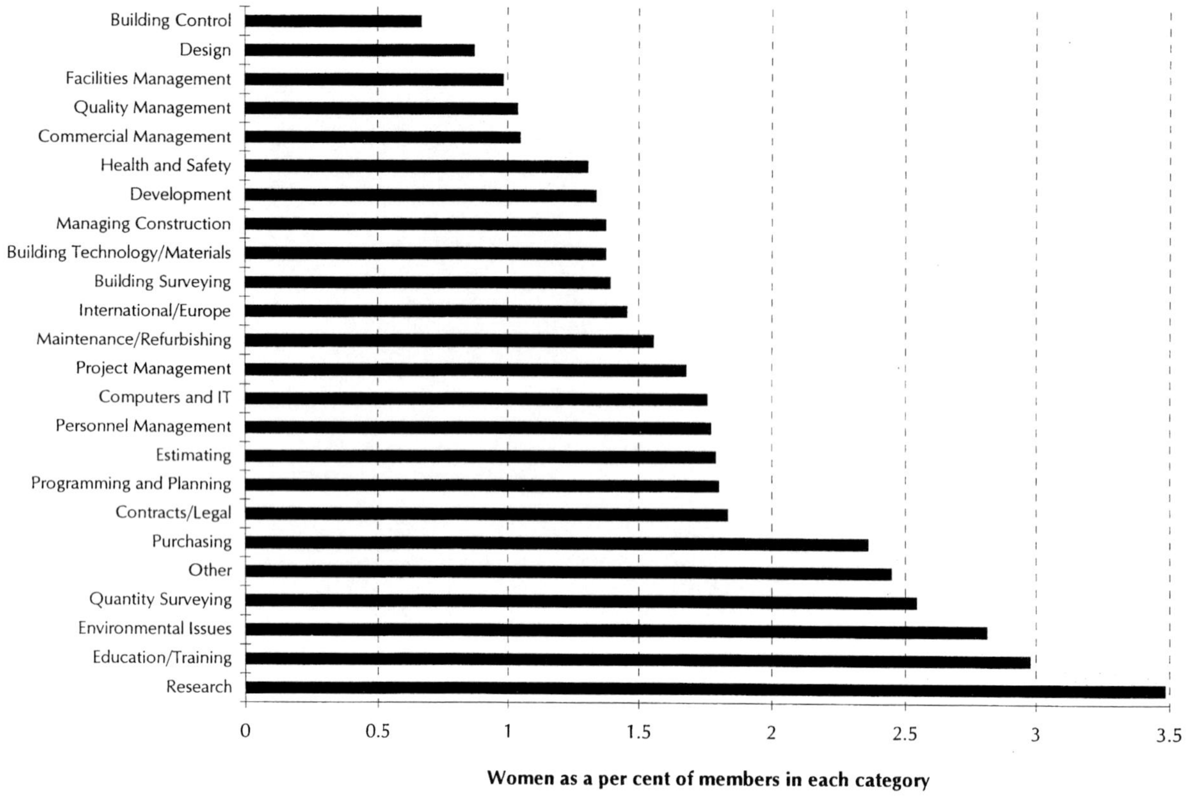

Building Control
Design
Facilities Management
Quality Management
Commercial Management
Health and Safety
Development
Managing Construction
Building Technology/Materials
Building Surveying
International/Europe
Maintenance/Refurbishing
Project Management
Computers and IT
Personnel Management
Estimating
Programming and Planning
Contracts/Legal
Purchasing
Other
Quantity Surveying
Environmental Issues
Education/Training
Research

0 0.5 1 1.5 2 2.5 3 3.5

Women as a per cent of members in each category

Source: CIOB membership survey, 1994

issues, quantity surveying) (Figure 2.2).[1] In addition, relative to their share of total CIOB membership, women are over-represented in public sector activities, including Housing Associations, Local and Central Government (Figure 2.3).

The issue of women's employment in construction does not, however, stop at their general under-representation within technical activities. There is also evidence to suggest that once they enter the industry, they do not progress at the same rate, or to the same extent as their male colleagues.[2] Indeed, a high proportion of women are aware of being disadvantaged at work because of their gender. In one study of the building industry,

1 Greed (1990a) highlights a similar situation in surveying. There are exceptionally few women in technical areas of surveying such as mineral surveying, but in the quasi-technical areas such as quantity surveying around three per cent of practitioners are women. Even here, however, many are involved in the legal side and not directly in construction.

2 This phenomenon is not just confined to construction or technical activities — it is a general issue for women in most occupations. Women are under-represented at the top of a range of organisations and professions, including the National Health Service (management and doctors), the Civil Service, personnel, banking, and general management (see Hansard Society Commission, 1990).

Figure 2.3 Type of employment of CIOB members: women as a per cent of total membership

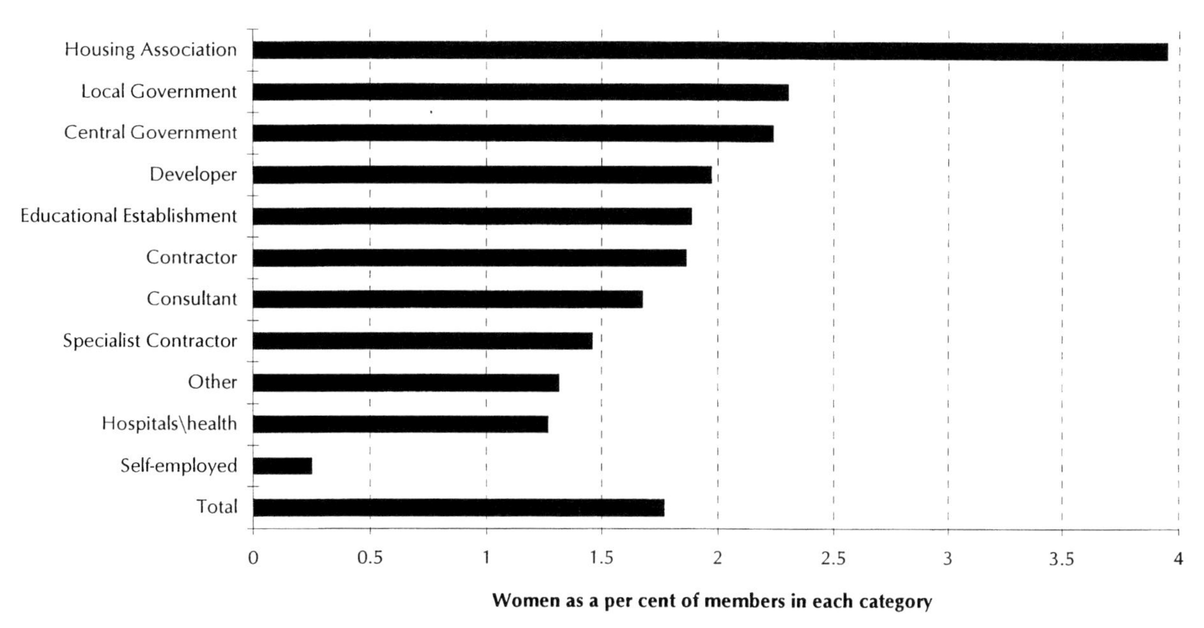

Source: CIOB membership survey 1994

for example, 48 per cent of women indicated that they had suffered discrimination at work. Furthermore, many of those who did not think they had been treated differently went on to describe incidents which clearly indicated the opposite (Lowe and Byrne, 1993). Likewise, over a half of women architects surveyed in 1992 felt they had experienced a situation where they had not been treated on equal terms because of their gender (Kirk-Walker, 1993). This may go some way toward explaining why so few women reach the top of their profession or even believe that this is this is an achievable goal (Davis, 1994; Millet, 1994).

2.3.2 Women in professional bodies

A second indication of the extent of women's representation in the construction professions can be gleaned from professional body membership data. These have been highlighted in a number of recent studies (Sommerville, Kennedy and Orr, 1993; Lowe and Byrne, 1993; Greed, 1994; Gale, 1991a) but it is worth pointing out that while women remain under-represented in all the relevant organisations, the situation is more extreme in some than others (Table 2.4).

Taking full members first, women are particularly poorly represented in the Chartered Institute of Building and the Institution of Mechanical Engineers. They form a higher proportion of members of the Royal Town Planning Institute, the Royal Institute of British Architects and the Royal Institution of Chartered Surveyors. There is a distinct pattern here, with the

Table 2.4 Women's representation in professional bodies 1993

	Full members %	Student members %	Total members %
Chartered Institute of Building	0.7	4.3	1.7
Royal Institution of Chartered Surveyors	5.2	15.2	7.6
Institution of Civil Engineers	1.0	10.0	3.5
Institution of Structural Engineers	1.5	11.3	4.2
Architects and Surveyors Institute[1]	1.1	4.3	1.2
Incorporated Society of Valuers and Auctioneers	5.6	17.0	8.5
Royal Institute of British Architects	7.0	25.0	7.5
Royal Town Planning Association	18.1	42.3	22.5
Chartered Institution of Building Services Engineers	1.2	4.9	1.8
Institution of Mechanical Engineers[1]	0.7	7.7	2.5

Source: Greed, 1994; Institution of Mechanical Engineers Membership Data Officer

[1] Data refer to January 1994

more technically oriented professions showing a greater degree of gender imbalance.

The second point evident from this table is that data on student members show a different pattern from that for full members. In each case, women account for a greater proportion of total student members than full members. This suggests that young women are entering technical professions in larger numbers than in the past. Data for the building industry bear this out. Figure 2.4 shows that the number of women in the CIOB has increased steadily over the past decade. More women are entering building professions but, since this is a fairly recent trend, they are younger than the average and therefore less likely to have reached full member status.[1]

Once age is taken into account, the differences in status disappear. Among the 30-40 year old age group, a slightly higher proportion of women have reached member or fellow status within the CIOB than is the case among their male counterparts (a third have compared to 30 per cent of men). This suggests that given the opportunity to enter the industry and then remain within it, women's progress, as measured by status within the professional body, is comparable to that for men.

[1] CIOB membership data indicate that 61 per cent of women members are 30 years old or under and 31 per cent are aged between 30 and 40. Among men these proportions are 19 and 28 per cent respectively.

Figure 2.4 Women's membership of CIOB: 1986-1994

Source: Lowe and Byrne, 1993, CIOB Membership Data 1994

Summary 2

This section has provided the context for the rest of the report by looking at trends in women's employment. The key issues highlighted include:

- the rise in the number of women in paid work, which means that they are projected to account for four-fifths of the growth in the labour force between 1994 and 2000.
- the increasing number and proportion of mothers in the labour market, and the implications of this for employment practices
- the large-scale entry of women into a range of previously male dominated professions, including for example accountancy and banking
- their continued under-representation in technology and science, especially the construction industry, despite recent growth in the number of women entrants to these activities
- the particular shortage of women at the top of technology and science oriented organisations and professions, including construction.

The next chapter looks at some of the reasons for these patterns by focusing on the barriers to women's entry into, and progress within, the construction industry and related activities.

3. Barriers to Entry and Progress: Construction and Related Industries

This chapter focuses on some of the reasons why women are under-represented in the construction industry by looking at the barriers they have to overcome in order to enter and then progress within it. It is primarily based on the findings of previous research, but we have also included relevant comments from the women surveyed and interviewed as part of this study.

Where possible we have summarised the results of studies on building or related activities. Analyses of other technically oriented, male dominated industries, in particular engineering, are included where they substantiate points which appear relevant but have not been studied within a building industry context. In fact, the issues are remarkably consistent across widely differing types of industry, with women experiencing similar challenges in a range of managerial or professional roles irrespective of the industrial activity concerned.

While not all women feel that they have been prevented from achieving their full career potential or treated any differently to their male colleagues (Evetts, 1994), the patterns identified in Chapter 2 are sufficiently consistent to have provoked the interest of researchers. The reasons for women's poor representation within construction and related industries have been investigated in a number of studies (Gale, 1989b; 1991a; Wilkinson, 1990; 1992b; Lowe and Byrne, 1993; Sommerville, Kennedy and Orr, 1993; Devine, 1992a; Corcoran-Nantes and Roberts, 1995; Evetts, 1994). These indicate that there are clearly two dimensions to the issue: women's apparent reluctance to enter these types of industry in the first place; and, if they overcome this initial hurdle, their experiences and opportunities for progression within them.

This chapter is divided into four main parts: the first looks at why women do not enter technical occupations, and the second summarises the challenges they face within them. This is followed by a short section on the advantages of being female in the construction industry. The final section examines a series of initiatives undertaken by the engineering industry in its attempt to increase the representation of women among engineers.

3.1 Barriers to entry

This section of the report examines some of the reasons for women's under-representation in technical occupations by looking at young women's educational and career choices. The relative importance of these two issues is difficult to disentangle because career choices, especially in technology and science, are frequently constrained by previous educational decisions, while, to some extent at least, educational decisions are based on some notion of an individual's potential career path. Whatever the underlying cause of their choices, however, the subjects young women study have a clear impact on their future access to technical occupations. Their educational attainment is therefore a key issue and is addressed separately from that of wider career choices.

3.1.1 Gender and technology and science qualifications

As noted in Chapter 2 above, the overall educational achievements of girls are rapidly approaching parity with, and at some stages surpassing, those of boys. Closer examination of recent trends, however, shows that at key stages in their lives, young women make different choices to their male counter-parts.[1] The difference between the two groups is no longer primarily one of *access* to educational opportunities, but revolves around the *direction* taken within the educational system. In this section, four key aspects of educational choice are discussed:

- patterns of GCSE attainment
- 'A' level science
- post GCSE vocational awards
- higher education decisions.

Patterns of GCSE attainment

Until science was made compulsory to age 16, girls' participation in science subjects at what is now GCSE level (formerly 'O' level) was low relative to that of boys.[2] In 1980, only 11 per cent of girls took physics compared to 31 per cent of boys. The introduction of GCSEs and the National Curriculum has changed this situation quite substantially. By 1993, the proportion of girls taking physics GCSE (including double-award science, which

---

1 The causes of this process lie considerably further back in a person's development but are manifested in the teens and early 20s because of the structure of the UK education system (see Section 3.1.2 for a discussion of the early development of gender specific behaviour).

2 The main exception to this is biology, in which girls formed the majority of candidates in both 1980 and 1993 (Smithers and Robinson, 1994, Chart 1.3).

includes a physics component) had risen to 65 per cent. The gap between boys and girls had also narrowed to just five percentage points, with 70 per cent of boys taking physics GCSE compared to 65 per cent of girls (Smithers and Robinson, 1994, Chart 1.3).

These patterns suggest that young women are not less suited to science than young men. When encouraged to do science subjects they prove more than capable. This is reflected in science GCSE grades, which show girls and boys gaining a similar proportion of good grades (A-C). For example, in 1993, 49 per cent of girls and 48 per cent of boys obtained grades A-C in double award science; in physics girls did better than boys, with 72 per cent gaining grades A-C compared to 67 per cent of boys (Smithers and Robinson, 1994, Chart 1.4).

'A' level science

The transformation in young women's access to science subjects at age 16 has not yet, however, carried forward into 'A' levels. Indeed, *fewer* young people are taking only science subjects and mathematics at 'A' level, although there has been a rise in the number taking a mixture of science and/or maths in conjunction with other subjects. Moreover, the decline has been as marked among young women as young men (in the case of physics, more so) (Smithers and Robinson, 1994, p.10-11).

This means that there has been little change in the proportion of technology, mathematics and science 'A' level candidates who are women (Figure 3.1). In 1993, women accounted for 54 per cent of all those taking 'A' levels but just 22 per cent of physics candidates and 18 per cent of technology candidates. They did,

Figure 3.1 Women as a proportion of all 'A' level candidates in 1992/3: selected subjects

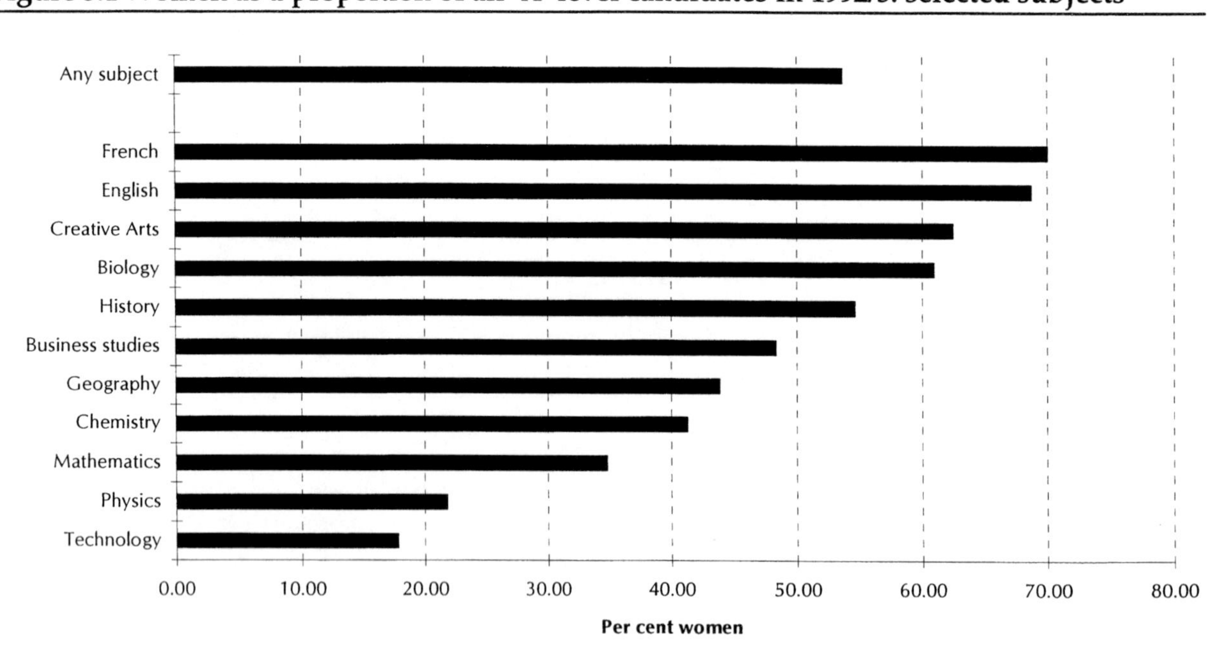

Source: DfE Statistical Bulletin, 1994a, 7/94, Table 13

however, form a somewhat higher proportion of those taking mathematics 'A' level (35 per cent), although even here their representation was low relative to subjects such as French and English (Figure 3.1).

Post GCSE vocational awards

The under-representation of women is even more marked in the vocational awards related to construction. Data from 32 colleges offering a GNVQ in Construction and the Built Environment shows that at intermediate level just three per cent of students were women in 1993/4, a proportion which doubled to six per cent at the advanced level (Sims, 1994).

A similar pattern is evident for other technology related vocational subjects. Considerably fewer girls than boys surveyed in 1990 had obtained or were seeking a vocational qualification[1] in engineering or architecture, building and planning, while the reverse was true in business and administration (Courtenay and McAleese, 1994, Table 5.6 and 5.7).

Higher education

While post-GCSE awards in technical subjects are not necessarily required for entry to a construction or building higher education program, an interest in such subjects is. It is not, therefore, surprising that women are also under-represented on higher education science and technology courses (Figure 3.2). For example, they account for only 15 per cent of full-time students on engineering and technology first degree courses and 22 per cent of those on architecture, building and planning courses. A similar pattern is evident for 'other undergraduate' courses[2] where women account for 13 per cent of full-time students in engineering and technology and 11 per cent of those on architecture, building and planning courses.[3]

Likewise, in 1992, 57 per cent of further education students were female, but just nine per cent of those on engineering and technology courses and five per cent on architecture, building and planning courses (DfE Statistical Bulletin, 1993).

---

[1] It is not possible to assess what type of vocational qualification is involved here.

[2] 'Other undergraduate' higher education courses include higher national certificate and diploma courses, courses leading to a variety of professional qualifications, individual institutions' certificate and diploma courses and some courses of degree equivalent standard (DfE Statistical Bulletin, 1994b).

[3] For both first degree and 'other undergraduate' courses, similar patterns are evident among part-time students (Higher Education Statistics, 1994, Table 24).

Figure 3.2 Women as a proportion of all home full-time higher education students 1992/3: selected subjects

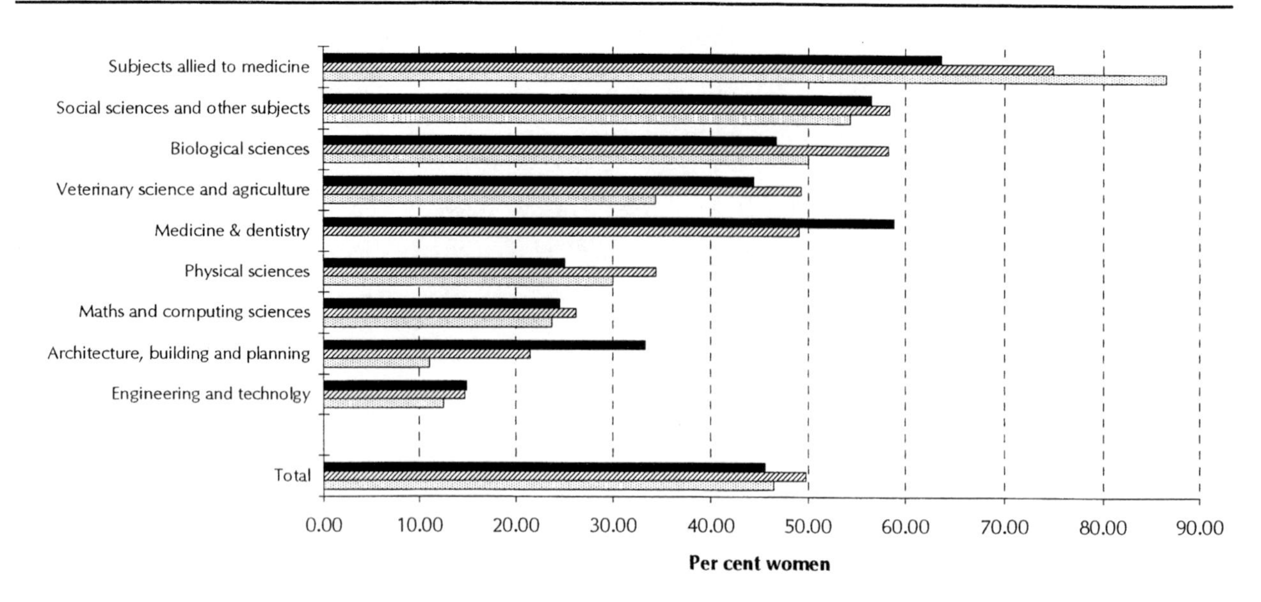

Source: DfE Statistical Bulletin, 13/94, Table 8

While these figures show that course enrolments remain highly gendered, the current situation is an improvement over previous years. The number of women among technology and science first degree graduates has increased over the past decade, as has their share of qualifications awarded in these subjects (Committee on Women in Science, Engineering and Technology, 1994).

This is reflected in data on civil engineering, building and architecture university first degree graduates. Table 3.1 shows that women's share of first degrees awarded in these subjects increased in the late 1980s and early 1990s, in the case of building from seven to eleven per cent.[1] In addition, 1993/4 data on all full-time students enrolled on these courses shows that in civil engineering and building the proportion of women is continuing to increase.

Similar data for the new universities show the same pattern, although women form a smaller share of the total. Women received 10 per cent of full-time and sandwich first degrees in building from the former polytechnics in 1992 (Gale, 1989b; PCAS, 1994; CSU, 1992). In addition, in 1986, seven per cent of admissions to building degree courses in these institutions were women, a proportion which had increased to over 10 per cent by 1992-3.

---

1 A similar picture is evident in other building related subjects such as building services, quantity surveying and building surveying (Gale, 1991a).

Table 3.1 University first degree graduates and students in selected subjects: UK domicile

	1985/6			1992/3			1993/4
	Men	Women	% women	Men	Women	% women	All students: % women
Civil engineering	903	91	9.2	1035	184	15.1	15.8
Building	179	13	6.8	412	51	11.0	12.1
Architecture	313	114	26.7	301	152	33.6	29.9

Source: USR, 1987 and 1994a and 1994b. Data refer to old universities only and exclude the new universities.

3.1.2 Choosing a career: influences and constraints

The educational decisions made relatively early on in life are clearly one of the factors which limit the employment opportunities open to women. An additional issue in any analysis of the under-representation of women in technology and science must, however, be the reasons why young women do not choose to pursue a career in these occupations.

This is itself an immense and complex topic and in this section we will only highlight the main findings of some recent research of relevance to this report. This shows that family, friends and other individuals are key influences on career choice or, at least, important sources of information on career options (Harris, 1988a; Hodkinson, 1994). For example, among 12-14 year old young men surveyed for the Construction Industry Training Board in the late 1980s, parents and careers teachers and advisers were most often mentioned as the people they took most notice of when deciding on a career. Women aged 18-24 identified similar influences, with most notice being taken of parents (especially mothers), careers teachers and careers advisers (Harris, 1988a, p.14-15).

A key issue for technology, engineering and science based industries is the kind of advice these key individuals are passing on to young people, and the information or perceptions on which it is based. When making career decisions young people and their advisers assess a number of factors. Important attributes for a potential career include (Harris, 1988a):

- a career's ability to deliver financial and job security, with this being of equal importance to males and females
- variety of work, with more young women highlighting this as an important feature
- respect from family and friends, which is again more important for young women
- team working, with this being of particular concern to women
- the opportunity to start their own business, which is more important for men.

These concerns suggest that one of the key factors influencing career decisions is the extent and type of knowledge about potential occupations or industries (broadly speaking, their image). Other relevant factors include:

- the perceived attitude of the industry to women
- the labour market prospects and conditions within different types of work
- the individual's aspirations and confidence (Rees, 1994).

Each of these issues is considered in more depth below.

Image

Individuals choosing a career, rarely have full information on the choices available. In the absence of countervailing factors such as personal or family experience, they are therefore heavily influenced by the image of the various careers or activities on offer. For industries such as engineering and construction this image can be problematic and has been identified as such in numerous reports (for example, Sommerville, Kennedy and Orr, 1993; Gale, 1991a; Greed, 1990a).

Research conducted for the CITB, for example, shows that the construction industry is strongly associated with lower status, manual work such as bricklaying, painting and decorating, carpentry, or plumbing, although among sixth formers and undergraduates there was greater awareness of professional activities such as engineering and architecture. Amongst this group, however, the status of construction as a career did not compare favourably with the other options available (Harris, 1989).

The main positive aspects of the construction industry were that there is a visible product of your work, the work is out of doors and varied, and pay levels. On the negative side, both women and men identified safety concerns and that the work involves getting dirty, manual labour and being out in poor weather (Harris, 1988a; Harris, 1989). The association with heavy, dangerous, manual work is a particular factor likely to reduce the industry's attractiveness to women.

The image 'problem' facing construction is a general issue which affects both young men and women. It is by no means only a factor in attracting women to the industry. Nevertheless, women are likely to be particularly affected. As highlighted above, young people are concerned about the effect of their career choices on the respect they will gain from their family and friends. Construction is a highly gendered activity. This is a positive attraction for young women who want to work in a less traditional job. For the majority, however, breaking with tradition can be hugely problematic. Research into attitudes in adolescence and vocational choice shows that the mainstream of young people will make choices which are seen as normal for their background (Byrne, 1992; Hodkinson, 1994; see also Gale, 1992).[1] What is seen as normal is of course highly and strongly gendered (Cockburn, 1987).[2]

There are good reasons for young people's reluctance to deviate from the 'norm': those who make gender atypical decisions often face hostility from their peers. For example, girls who chose engineering YTS placements were simultaneously accused of being 'boy crazy' and 'lezzies' (Rees, 1994).[3]

It is not, therefore, surprising that a continual refrain in research on women entering non-traditional occupations is that they received strong encouragement and support from some quarter — their parents, a teacher or careers adviser — which enabled them to overcome the negative reaction their choice provoked elsewhere (Greed, 1990b; Devine, 1992a; Carter and Kirkup, 1990). For example, Greed has this to say about the influences on women surveyors:

> *'Some girls are being told by women teachers that they are 'naughty' or 'going through a rebellious phase' for wanting to be a surveyor. Indeed, many depended on the support of their parents against the school in order to be allowed to do the right 'A' levels and apply for surveying courses.' (Greed, 1990b)*

---

[1] What is viewed as 'normal' develops very early. By age 11, if not before, both boys and girls have developed quite strong sex-stereotyped attitudes concerning the roles of women and men in society and appropriate behaviour for boys and girls. Even primary school children have firm views on these issues (Chivers, 1986; Smithers and Zientek, 1991). Technical and scientific activity in the UK remains strongly associated with men, although with science now compulsory in the National Curriculum and more women entering non-traditional jobs this may slowly change.

[2] The strength of job gendering is such that young women find it difficult to conceive of being a woman in some jobs. For example, girls who met women role models in banking (a bank manager) and the airline industry (a pilot), six months later remembered them as a bank clerk and an airline stewardess! A chemical engineer found that she was repeatedly asked what she did, and when questioned as to why, the girls concerned said it was because she had a handbag. Evidently the combination of engineer and handbag was something they found difficult to accommodate (Rees, 1994).

[3] The decline in single-sex education within the state sector may be a factor here — girls are more likely to study science in single-sex schools or classrooms (Rees, 1994; see also Devine, 1992b; Stone, 1992). One of the reasons that girls in single sex schools find it easier to pursue interests in science and technology may be due to their relative immunity from attacks by fellow pupils for choosing 'male' courses during a difficult time for personal development, adolescence. Stone (1992) notes that a high proportion of the women engineers she is acquainted with went to single sex schools.

A second factor related to image, concerns the level of knowledge about careers in construction. If the industry is identified primarily with manual work, young people seeking a more professional role will not think of a career in construction. As well as being influenced by more gender neutral reasons such as poor status or job insecurity, women may not be made sufficiently aware of the wide range of jobs, including non-manual work, available within the industry, including non-manual (Srivastava, 1992).[1] Two of the respondents to the survey conducted as part of this study summarised this issue as follows:

> *'I feel there are few women in this profession because they do not know what jobs exist and what they could be capable of. I would have been a teacher or a nurse if it had been left up to my school. I was greatly helped by my father, himself a builder, who gave me insight and initial contacts for employment.' (Building Surveyor)*

> *'The main reason why there are not many women in the industry is due to the general lack of information on careers available. Women with good qualifications usually are pointed in the direction of accounting* etc. *No information on the range of jobs/careers available.' (Quantity Surveyor)*

This lack of knowledge has been identified as one of the main reasons why women do not enter construction careers (Gale, 1991a).

Alternatively, women who do wish to pursue a career in the industry may find themselves being channelled away from construction altogether, or into particular parts of the industry, because of the image of the jobs involved or the lack of role models within them (Lowe and Byrne, 1993; Srivastava, 1992). This was confirmed in the survey, with one woman noting:

> *'There is definitely a lack of females in the industry and this is in the main due to the public image as well as the discouraging attitudes of teachers at school. I myself was not allowed to choose 'O' level Building Studies since the deputy headmaster thought that I would not 'suit' the subject.' (Construction Manager)*

Perceived attitude of the industry to women

Among young women and their parents, construction is viewed as having a poor attitude toward the employment of women. A study conducted in the late 1980s asked young women and their parents about their views on the industry. Both groups thought that women would find it more difficult to get a job in the industry, would not receive equal treatment at work, would

[1] Informal recruitment practices do not help here. These are likely to have less of an impact on women looking for professional careers but even here they act to limit young women's ability to find out about openings in the industry.

suffer sexual harassment and be discriminated against when it came to access to the best jobs (Harris, 1988a). To a large extent, these views appeared justified. The majority of employers, when asked about their views, thought that women were less able than men to work satisfactorily in construction and that there were a lot of employers (and clients) who would be unhappy to employ them (Harris, 1988b).

This perception of the construction industry is compounded by the lack of role models to challenge the dominant view. While women are under-represented in construction, they are not entirely absent. This is a factor which women with knowledge of the industry are likely to be aware of but which needs to be communicated to a wider audience if the current gender balance is to change.

Labour market prospects and conditions

As we have seen, young people and their parents are concerned about the ability of a career to deliver financial security and a reasonable prospect of employment. Young women, however, are also influenced by the extent to which a potential career will allow them some degree of flexibility when combining home and working lives (Rees, 1994). The brighter students in particular do not expect to leave the labour market when they have children, but they are realistic about persuading their future partners to share the domestic and childrearing workload on an equal basis. This remains one of the reasons why among more qualified young women occupations such as teaching, nursing, and pharmacy continue to be attractive options. For example, a study of the degree and career choices of young people qualified in science subjects found that:

> *'... female students in several cases had already considered the issue of combining an S/T[1] career with having a family. Their perception was that S/T employers, particularly from the private sector, have reacted more slowly to calls for career breaks and childcare support for working women than have some other professions such as banking. In some cases, this had contributed to girls' decisions to opt for non S/T degrees and for other S/T students it was an influence on their career plans.' (Fuller, 1991, p.338-9)*

Low aspirations and confidence

A final factor influencing women's career choices is their aspirations and confidence. This is related to the issues around the gendering of jobs and the perceived labour market prospects discussed above. There is, however, an additional issue not yet covered. This is the impact of low occupational aspirations and

1 Science or technology.

low levels of confidence among women, in particular those from working class backgrounds (Rees, 1994).

This is important for the purposes of this study because of its implications for skills development in later life. Many women are quite able to engage in technical roles but have not chosen to do so sufficiently early on in their lives to gain the right qualifications and necessary experience (Srivastava, 1992). When making career decisions, they will have been influenced by the kinds of factors already discussed and not had the confidence or awareness to consider a non-traditional role.[1] Alternatively, they may have selected a career path which does not make full use of their potential. As a result, there may be a large number of women already working in the construction industry who, with some encouragement and appropriate training, could be channelled into more demanding, technically or managerially oriented positions.

3.2 Barriers to career progress

Despite these barriers to entry, the number of women choosing to follow a technical or professional career in construction and related activities is increasing. This section looks at the experiences of those who have elected to work in a technology related field and identifies the barriers they encounter to remaining, and progressing, in their chosen career. There are two main categories of issues here. The first concerns the difficulties of balancing the demands of a career with those of family life, including assumptions about women seeking to do this. The second relates to a range of attitudinal and cultural issues which inhibit women's progress and retention within building. Each of these is discussed in more detail below.

3.2.1 Combining work and family life

'Women, like men, should be free to have or not to have children as well as worthwhile careers.' (H Rose, quoted in Patel, 1994)

'Childcare responsibilities were judged the greatest impediment to career progression.' (Devine, 1992a)

'The structure of the industry does not lend itself to family life — there is a culture of extremely long working hours and I would

---

[1] The education and training systems themselves have been identified as major perpetuators of occupational segregation (Deem, 1984; Cockburn, 1987). Certainly, despite a range of initiatives, the youth training system has had limited success in altering traditional patterns. In 1990 the largest occupational groups for women on YTS (as it was then called) were catering, cleaning and personal service work (37 per cent) and clerical and related work (10 per cent) (Courtenay and McAleese, 1994).

imagine women with children find it impossible to combine a 12 hours working day with childcare. Success and one's worth to the company are often measured by the hours put in rather than the quality of work. I believe that the industry pays only lip service to encouraging women. No structural changes, specific encouragement or help is given. As such it can only offer career satisfaction, development and progression to younger childless women. Having reached a stage where I want to start a family, I am having to consider leaving the industry — something I would rather not do.' (Facilities Manager)

'"Lots of assumptions are made that don't necessarily tally with the truth". Like the one that if you leave the office at 6 o'clock to get home to your kids, you can't be as hard working as someone who stays on every night until eight or nine. "It should be considered whether a very effective, hard-working woman can accomplish more in a shorter day than perhaps someone who spins out their hours unthinkingly."' (Chevin, 1994)

The *Rising Tide* report points out that the most common child-rearing period, when parents are aged between 25 and 35, coincides with the time of key development in an engineering or scientific career (Committee on Women in Science, Engineering and Technology, 1994). Likewise, in many areas of construction this is a key period for gaining the kind of experience which qualifies an individual for more senior posts. In short, it is when an individual's career becomes established.

Most of the responsibility for parenting in the UK remains with the mother, and it is women who are likely to have to take time off for childrearing purposes (in addition to, of course, childbearing). Knowing this, many women in technology and science (and, indeed, many other occupations) feel that they have to choose between their career and motherhood (Wilkinson, 1990; 1992b; Committee on Women in Science, Engineering and Technology, 1994). If they have children, they are generally faced with two options. They can either stop altogether or accept that their career will be sidelined as long as they have responsibility for children. Their male colleagues do not usually face a similar choice.

There are two issues here. First, the actual practical difficulties of combining work and family, and employers' attitudes toward, or assumptions about, women with children. Taking practical difficulties first, the lack of affordable, flexible childcare and nursery provision means that it is physically difficult for women to remain in work full-time (should they want to) while taking care of children (Committee on Women in Science, Engineering and Technology, 1994; Devine, 1992a; EOR, 1990a/b; Wilkinson, 1992b). This difficulty is compounded by the scarcity of part-time employment opportunities and career break schemes, in particular for those in supervisory or managerial roles (Kirk-Walker, 1993; Wilkinson, 1990).

The second issue is that of employers' attitudes toward women with childcare responsibilities. One study of women in engineering and science professions showed that women who did have children found their opportunities for advancement blocked. They were no longer deemed capable of fulfilling the requirements of higher positions (Devine, 1992a; see also Evetts, 1994). Managers assumed that *all* women were inevitably transformed by having a family, becoming less committed to their careers and no longer interested in having a demanding job (Devine, 1992a). They did not make the same assumptions about men.

Evidence from the construction industry points to an even more severe attitude. In one study, 21 per cent of employers indicated that having a family ends a career for women in the industry and 48 per cent thought that they became more difficult to employ because of restricted mobility (Wilkinson, 1992b; see also Stone, 1994; Chevin, 1994; Greed, 1990c). This was echoed by one of the respondents to the survey who pointed out that:

> *'On a personal basis I don't feel that I get a raw deal. However, I think trying to combine my job and children would be almost impossible and would* definitely *be frowned upon within the company I work for where I am the only senior female manager.' (Quantity Surveyor)*

The key issue here is not whether mothers do, or do not, wish to continue working in the same job they held prior to having children. Rather, it is the assumptions managers make which rob women of the opportunity to make these decisions for themselves.

3.2.2 Cultural attitudes and organisational norms

The difficulties of combining a career and family are underpinned by a series of cultural issues which affect everybody working within the industry but have particular implications for women. These factors are by no means confined to building, but the characteristics of the industry mean that some are of especial relevance to it. The main issues here include:

- assumptions about women's career aspirations
- treating women differently
- unease about women in positions of authority
- women and men in organisations.

Career aspirations

> *'Some of them think its only a hobby until you get married.' (Corcoran-Nantes and Roberts, 1995)*

> *'"Asian women will be married off in a few years, so what's the point of training them?" are comments I hear time and time again.' (Construction Manager)*

'I also feel the greatest deterrent to women in the Building Industry is that as soon as they get engaged or married, all training or career prospects are stopped because SENIOR MANAGEMENT are just waiting for you to leave to have kids.' (Construction Manager)

These comments suggest that even before women have children or get married, the possibility that they might do so makes managers view them as a risk or potential problem, especially if they occupy a senior position. To counter such views one of the participants in the discussion groups pointed out that the project based nature of many jobs in construction makes them quite suited to women who wish to take time off for family reasons and are prepared to plan their break from work to fit in with the timing of projects.

In construction, an indication of the prevalence of these attitudes is the number of women who report being asked in interviews about their personal lives and whether they planned to have children (see Greed, 1990c; Bolton, 1994). Some of the women we spoke to had gone to great lengths to reassure employers that they had no intention of having children despite the fact that they were married.

These assumptions contribute to others about the career aspirations of women, for example, that they are not ambitious or will not want to move around the country. Even recent studies report that women have to *demonstrate* their interest in career progression, whereas in men such aspirations tend to be assumed (*eg* Corcoran-Nantes and Roberts, 1995; Hirsh and Jackson, 1990). This makes a difference because if an individual is expected to progress, they will be offered the opportunities required for advancement, whereas their less fortunate colleagues will have to demand the same opportunities. In a culture which is highly ambivalent about assertiveness from women, this can be problematic.

Different treatment

'Many employers do not take women in the building industry seriously, therefore women have to work twice as hard and have to prove themselves.' (22 year old Building Technologist)

'In my five years within the industry I have found that equal opportunities doesn't work! I was training as a QS and had good qualifications but very poor pay whereas a male with less qualifications and training received more pay — I wonder why I left!!!' (Former Quantity Surveyor)

'In general, I have to prove myself to gain respect at this level, in comparison to colleagues who are taken at face value.' (Project Manager)

'In my experience I have found that when men meet men in the industry, each expect the other to have building knowledge and experience. However, when men meet women in the industry, the

women are expected not *to have building knowledge and experience. This can be very frustrating.' (34 year old Maintenance/ refurbishment professional)*

'One of the biggest things that has discouraged me from attending CIOB dinners is that women are invited as 'guests only'! or so it appears to me. Do they not think it might be nice to say dinner jackets/cocktail dresses for dress option? Don't forget women builders have partners too!' (30 year old Maintenance/refurbishment professional)

As well as having their career aspirations called into question, women are treated differently from their male colleagues in other respects. The issues which have been identified here include:

- the difficulty many women have in being taken seriously as professionals, and problems in dealing with the patronising, even hostile, approach of some managers or co-workers[1] (Kirk-Walker, 1994; Corcoran-Nantes and Roberts, 1995; Carter and Kirkup, 1990). For example, women find that their suggestions are ignored or dismissed by co-workers, whereas the same idea is embraced when introduced later by a male colleague (Yates, 1992). In some fields this issue may be compounded by the notion that women do not belong in the occupation or industry they are seeking to progress within.
- the tendency for managers to 'sponsor' or 'mentor' male employees, giving them advice and encouragement, integrating them into important informal networks, providing them with appropriate experience for promotion, and advertising their skills to other managers. Women, on the other hand, tend to be left to get on with their own careers without such invaluable guidance (Byrne, 1992; Devine, 1992a; Carter and Kirkup, 1990).
- dual standards in promotion decisions which make women suspect that they have to be better than their male colleagues in order to progress as rapidly or even be paid the same (Wilkinson, 1992b; Devine, 1992a).
- informal interactions within organisations, including professional bodies. These allow colleagues to establish good relations with each other and help to integrate individuals within organisational and professional structures. These informal interactions often occur in places and times which limit participation by women. They may take place, for example, at sporting events to which women are not invited or in particularly male dominated venues. In addition, women may feel excluded by the assumption that the target

[1] As one respondent in this study put it: 'They refer to you as a young lady as if you've got a disease.' (Corcoran-Nantes and Roberts, 1995).

audience is male (eg invitations addressed to member and wife).

Women in positions of authority

> *'There is a general feeling that men are likely to be groomed for partnership earlier, whereas women, after an initial climb, are shunted horizontally into specialist areas rather than continuing vertically. Indeed, many men appear to see women as 'helpers' or 'assistants' rather than decision-makers in their own right.' (Greed, 1990c)*

> *'There are many males within the industry who feel women are incapable of managing predominantly male personnel. This is so wrong. With the correct management education I feel women are less confrontational/aggressive and can handle situations more diplomatically!' (31 year old Building Student)*

> *'Apart from a few early examples about 12 years ago I have not experienced any direct discrimination in the industry. However, I believe my career has been subject to indirect discrimination through lack of promotion...Many of the older managers seem to expect that the only women in the industry are secretaries!' (Facilities Manager)*

There remains a profound unease about women occupying positions of authority and a lack of confidence in their ability to do so, in particular when they will be supervising male employees (Wilkinson, 1992b; Savage, 1992; Corcoran-Nantes and Roberts, 1995; Devine, 1992a; Carter and Kirkup, 1990). The result is a reluctance to promote women into such positions. This means that women fail to get the experience required for movement into senior management. They find themselves channelled instead into more peripheral activities which emphasise technical or professional expertise rather than supervisory or managerial responsibilities and authority (Evetts, 1994; Crompton, 1994).

The most extreme example of this attitude is when women are assumed to be secretaries rather than professionals. At this point the unease about women in positions of authority comes very close to a general undermining of their worth in the workplace.

Women and men in organisations

There are some indications that women and men behave differently in organisations. As with other cultural issues, these patterns obviously have their roots in wider processes, and it is at this level which they need to be addressed, but they can have a marked effect on career progress.[1] Of particular importance here are issues about self-confidence, self-esteem, assertiveness,

---

[1] The issues discussed in the previous sections obviously have a role to play here. If women find their working environment undermines or devalues them, they are unlikely to be as self-confident or aware of their abilities as their male colleagues.

and the capacity for self-promotion (eg Matthews, 1994) In engineering and science occupations, as is the case elsewhere, career progress is heavily determined by managers' views of an individual's work and potential for development. Avoidance of taking personal credit for successful projects or failing to bring that success to a manager's attention are not, therefore, optimum strategies. Likewise, having the confidence to go for promotion or ask for a pay rise can make a substantial difference to the speed at which an individual moves up the organisational hierarchy.

Many of the women we spoke to in the course of this project described themselves as having become more assertive as a result of their work. They had successfully developed strategies to overcome some of the negative aspects of female socialisation, including lack of confidence or inability to cope with a competitive working culture/management style. This had served them in good stead at work, although some did express reservations about its effect on relations with friends and family.[1] In the absence of changes to the working culture, for women new to the industry, advice or training on developing these kinds of strategies may help them to integrate better, or faster, into the working environment.

3.3 The advantages of being female

'The industry must increase the proportion of women in senior roles if it is to survive and beat the competition in the form of European contractors. 50% female representation may help to reduce the conflict and antagonistic attitudes which are so damaging to the industry.' (Quantity Surveyor)

This chapter has so far focused on the barriers to women's entry and progression within construction occupations. Alongside the issues discussed in Section 3.2, however, the more positive aspects of being female in a male dominated industry need to be highlighted. These include:

- visibility — this can obviously work both ways in that if a woman makes a mistake she is likely to be remembered for it. Women's achievements or contributions are, however, more likely to be remembered if they are the only female involved (Davis, 1994; Davies, 1987).
- ability to develop good relationships with clients and co-workers (Davis, 1994; Chevin, 1995). This can stand a woman in good stead in those areas of construction where

---

[1] This tension between the kind of behaviour valued at work and that expected by friends and family has also been highlighted by Carter and Kirkup, 1990 and Gale, 1992.

harmonious relationships with clients and colleagues are an advantage.

3.4 Women into engineering and science initiatives

This summary of existing research has focused mainly on the problems women face in ensuring that their careers progress at the same rate as those of their male colleagues. A key prior issue, however, is that of encouraging women to enter technical occupations in the first place. This and the various strategies designed to change the current situation are the focus of the next part of this chapter.

Over the past two decades there have been a range of initiatives to encourage women into science and technology careers (see Rees, 1994 for a summary). These include:

- Girls into Science and Technology project (GIST)
- Girls and Occupational Choice Project (GAOC)

- Women's Training Roadshows
- Engineering industry initiatives (*eg* GATE (Girls and Technical Education Project), Insight, EITB Technician Engineer Scholarship Scheme, WISE).

Of these the latter are probably the best known and, because of their particular relevance to this report, they are discussed in more detail below.

3.4.1 The WISE Campaign

The WISE campaign began in 1984 as a joint venture of The Engineering Council and the Equal Opportunities Commission. It was aimed at changing the attitudes of young people, parents, teachers and the general public about the suitability of engineering as a career for girls (EOR, 1990b).

The campaign has taken a variety of forms including:

- a publicity campaign in women's, teenagers' and educational publications
- the mailing of posters and leaflets to schools
- WISE buses — mobile teaching and exhibition centres in which groups of girls (aged about 13 and 14) spend time with tutors and gain hands on experience of technology
- a trailer sponsored by British Rail's Signal and Telecommunications department, and staffed by women technicians who visit schools
- a series of booklets detailing the various initiatives aimed at encouraging women to consider a career in engineering, and

providing information to parents, young people, teachers and lecturers.

The campaign is financed by sponsorship from outside bodies, including employers.

The WISE campaign has not been formally evaluated although funding is currently being sought for such an exercise. The initiative has, however, provoked interest. Some 800 enquires were received in the first year and an estimated 150,000 girls have participated in the WISE Vehicle Programme (the buses and trailer) (WISE Campaign, 1994).

3.4.2 EITB Initiatives

The EITB developed a series of initiatives aimed at attracting young women into engineering and encouraging firms to recruit them (EOR, 1990b). They include:

- Insight — one week residential courses for sixth formers. They are held at universities and designed to encourage women to take an engineering degree.[1]
- Taster — a one week non-residential course at EITB Training Centres and designed to encourage young women to apply for a Craft/Technician Training Award.
- GATE — regional courses of three to five days to encourage girls to consider a career as an engineering technician.
- TESS — a two year scholarship leading to an HND. TESS was designed to demonstrate to employers, women and careers advisers that women over the age of 18 with 'A' level potential can be recruited and trained to adult technician engineer level.

These initiatives began in the late 1970s in response to an anticipated skill shortage. With the recession and the demise of the EITB and its replacement by the more commercial Engineering Training Authority (EnTra), there have been substantial cutbacks and the only initiative still running on a national basis is Insight.

A review of the programmes carried out in 1987, however, showed that they were having a positive impact on individual participants.[2] For example, among Insight participants in the early 1980s there was an increase (from 40 to 50 per cent) in the

---

[1] In 1994, the Insight Programme was supplemented by Engineering Education Scheme courses at Durham, Oxford Brookes and Brunel.

[2] Insight type courses in the construction industry also received positive feedback from participants (Gale, 1991b; 1991c), although an ongoing formal programme of events has not been developed.

proportion considering an engineering de؍
participants the proportion considering e
from 37 to 30 per cent (EITB, 1987a). In th
per cent of young women taking part in t
subsequently opted for engineering degr
1990b).

TESS also provided a successful alternative route in
engineering for scheme participants. Almost all those recruited to courses between 1985 and 1987 completed the scheme and 81 per cent were successful in gaining an ONC or HNC. Moreover, 79 per cent found employment in engineering, with an additional seven per cent going on to engineering related higher education. The main limitations of the scheme were that not all TESS places were taken up, an indication of both the power of negative stereotypes and the need for more widespread publicity, and that it had limited impact on the engineering industry as a whole (Callender, Toye and Connor, 1992).

3.4.3 Evaluation issues

During the course of the WISE campaign and other engineering industry initiatives, the proportion of women on engineering courses has increased from seven per cent in 1984 to a current level of 15 per cent. The extent to which this can be attributed to the various campaigns is, however, unknown. Nevertheless, a number of key issues concerning the effectiveness of actions to encourage women to enter science and technology have been highlighted. These include:

- the extent to which the campaign really changes attitudes. Some young women in science and technology who had been exposed to the various initiatives felt that they tended to 'preach to the converted' rather than transforming the options open to all women (Fuller, 1991)
- a related issue — that while individual outcomes appear to be positive, the intervention may come too late in young women's careers, or be too limited in scope, to be effective on a wider scale (Fuller, 1991)
- that the initiatives are expensive and as such are unlikely to be made more widely available. Indeed, the current economic and training climate has been associated with a marked cutback in all but two of the programmes (WISE and Insight).
- the indications are that the image of industry remains an issue despite more than a decade of intervention (Evetts, 1994; Fuller, 1991; Meickle, 1995)
- there has been substantial activity to encourage women to enter engineering occupations but less has been done to encourage them to remain and progress within them. Some organisations participating in the WISE programme did develop equal opportunities initiatives within the workplace

(McRae, Devine, Lakey, 1991), but there has been no concerted industry wide campaign equivalent to that directed at overcoming barriers to entry

- finally, the initiatives have focused on changing the attitudes of women. There has been little activity designed to encourage more widespread change. The impact of this is as it relates to the problems of combining home and family life in technical careers is summed up in the following statement:

> *'Girls' assumptions about their future roles in the family, and what that leaves over for participation in the labour market, clearly has an overwhelming effect not simply on what choices are made, but whether choices are effectively made at all. Initiatives that are designed to encourage girls into fields currently the reserve of men have tended to ignore the other side of the equation. There have been no initiatives on the same scale designed to encourage boys to take a more active role in childcare.' (Rees, 1994)*

Summary 3

This chapter has looked at the main barriers to women's entry into and progress within construction and other male dominated industries. It has drawn on previous research in the field as well as comments from women participating in the CIOB study.

3a Barriers to entry

The main entry barriers include:

- the educational choices young women make, with the number choosing technical subjects at school and in postsecondary education remaining substantially below that of men
- the image of the construction industry, which means that it is often not viewed positively when women are making career decisions. This is compounded by the lack of awareness of the range of careers available within the industry
- the perceived attitude of the industry towards women
- labour market prospects, with some women being reluctant to pursue careers in industries which make it difficult to combine a career with family life
- young women's low aspirations and confidence. This means that they do not think of a technical or professional career in the first place and so do not gain relevant qualifications or experience.

3b Barriers to progress and retention

Progress and retention are affected by:

- the difficulty of combining work and family life and the assumptions made about women seeking to do so.
- assumptions about the career aspirations of women, in particular that they will leave to have a family or that their career is less important than that of their partner.
- the different treatment women receive, ranging from a lack of recognition of their professional expertise and technical competence to their exclusion from important informal interactions.
- unease about women holding positions of authority and the tendency to channel them into specialist rather than generalist positions within the construction industry.
- women's lack of confidence and hesitancy about claiming credit for their achievements.

3c Advantages of being female

The main advantages of being a woman in male dominated industries include:

- visibility — which means that achievements or contributions are remembered.
- the kinds of skills women can bring to the job, in particular their ability to form good relationships with clients.

3d Initiatives

Finally, we looked at a series of initiatives designed to improve women's representation in technology and science These have had a strong positive impact on the individuals involved. They have, however, been relatively limited in scope and have yet to transform broader societal attitudes. As a result, change remains slow, although the number of women studying technology and science has increased over the past decade.

4. Characteristics of Women in Building

In this, and the next three chapters, we report the results of the postal survey of women with experience in the building industry (see Appendix 1 for details of the survey). The discussion is divided into four main topics:

- the characteristics of women in the building industry
- the career decisions
- the views of students on building courses
- women's views on the industry and the effectiveness of initiatives to encourage more women to enter and pursue careers within it.

In this first chapter we look at the following topics:

- the personal characteristics of women with experience of the industry
- their educational and professional background
- their work and working arrangements.

4.1 Personal and educational characteristics

4.1.1 Personal characteristics

Tables 4.1 to 4.3 show that women with experience of the building industry are a distinct group when compared with all women in employment. The main points to emerge from these tables include:

- that women building professionals are younger than is the case for all women in paid work
- former members of CIOB are slightly older than current members
- the proportion of CIOB members and former members who are from minority ethnic groups is similar to that among all working women
- women in building are less likely to have children than all women in employment.

Table 4.1 Age distribution of women in building and all women in employment

	All women in survey	CIOB members	Former members	Other	All women in employment (1994)
Aged 16-24	39.9	39.4	33.0	48.9	16.0
Aged 25-34	44.2	43.7	53.6	35.6	27.0
Aged 35-44	11.4	12.5	11.3	7.8	26.0
Aged 45 and over	4.5	4.3	2.1	7.8	31.0
Total cases (=100%)	466	279	97	90	–

Source: IES Survey 1994; Labour Force Survey Spring 1994

Almost two-fifths of women in building were aged between 16 and 24, while an additional 44 per cent were aged 25-34. This compares with 16 and 27 per cent respectively for all women in paid work. Over a half of former members are aged 25-34, a figure which falls to 44 per cent among current members and 36 per cent in the 'Other' group (Table 4.1).

A total of 92 per cent of respondents to the survey were white women, although this overall figure is distorted by the particularly high percentage of members of minority ethnic groups among the 'Other' sample (Figure 4.2). Members and former members of CIOB show a distribution by ethnic group closer to the average for all women.

Only 16 per cent of women responding to the survey had dependent children and less than two per cent were caring for an adult (Table 4.3). A higher proportion of former members of CIOB had children, an indication that this may be one of the reasons for leaving the industry (see Chapter 5 for a more detailed discussion of reasons for leaving). Overall, the

Table 4.2 Women in building and all women in employment by ethnic group

	All women in survey	CIOB members	Former members	Other	All women in employment (1994)
White	92.3	95.0	96.9	78.7	96.0
Black-Caribbean	1.7	1.4	0.0	4.5	1.0
Black-African	2.6	0.7	0.0	11.2	–
Black-Other	0.2	0.0	0.0	1.1	–
Indian	0.2	0.4	0.0	0.0	1.0
Pakistani	0.6	0.4	1.0	1.1	–
Chinese	0.6	0.7	0.0	1.1	–
Other	1.7	1.4	2.1	2.2	1.0
Total cases (=100%)	466	280	97	89	–

Source: IES Survey 1994; Labour Force Survey Spring 1994

– indicates that no data are available.

Table 4.3 Women in building and all women in employment by whether have children

	All women in survey	CIOB members	Former members	Other	All women in work (1994)
No dependent children	83.8	85.8	76.3	85.6	63.3
With dependent children	16.2	14.2	23.7	14.4	36.7
Total cases (=100%)	468	281	97	90	–

Source: IES Survey 1994; Sly, 1994, Table 4

proportion of women in the survey with children is considerably lower than among all women in employment, over one third of whom are mothers. This difference can be explained partly by the lower age profile of women in building. Nevertheless, even once age is controlled for, by taking respondents aged 25-39 only, the proportion of women in building with children is lower. 23 per cent have children compared to 60 per cent of all women in employment (derived from Sly, 1993, Table 3).

Among women with children, two thirds had at least one child under five and two-fifths at least one aged 5-16. Similar figures for all women in employment show that of those with children, 38 per cent had children aged under five (Sly, 1993).

A half of respondents to the survey were married or living with a partner. Among all women in employment, this figure rises to 73 per cent, again reflecting the younger age profile of women in the building industry. Women in the industry may also feel that it is better to remain single if they wish to progress in their career.

Parental background

We also asked about the family background of women in the survey. Anecdotal evidence suggests that having a parent working in a similar area is one of the factors which help women overcome the barriers to entering technical occupations (see Chapter 3). In the survey, two thirds of respondents stated that either their mother or father (most often the latter) worked in a professional or technical capacity. Of these, forty-two per cent

Table 4.4 Family experience of the construction industry (row per cent)

	Mother only	Father only	Both	Neither	No.(=100%)
Parents work(ed) in a professional or technical capacity	2.6	40.9	23.7	32.9	465
Parents work(ed) as prof./tech in construction	1.6	37.9	2.9	57.6	309

Source: IES Survey 1994

Table 4.5 Academic qualifications

	Current highest academic qualification				Expected
	All	CIOB members	Former members	Other	All
Higher degree	4.3	3.1	6.7	5.9	7.3
First degree	36.6	40.1	33.3	29.4	14.5
'A' level or equivalent	26.8	28.2	30.0	18.8	1.7
GCSE	25.6	22.1	24.4	37.6	1.1
Other	4.6	4.2	5.6	4.7	1.9
No academic qualification	2.1	2.3	0.0	3.5	0.6
Total cases (=100%)	437	262	90	85	468

Source: IES Survey 1994

[1] These data are based on the number indicating they expected to gain each qualification. It is not possible to distinguish between those who did not expect to gain the qualifications and those who did not answer the question. As a result, the column does not add to 100%.

had one or more parents working in the construction industry (Table 4.4). That is, over a quarter of women with experience of the building industry had a parent who had worked within construction in a professional or technical capacity. A comparable figure for the population as a whole is not available, but in 1991 less than one per cent of all those in employment[1] worked in a professional or technical capacity within the construction industry. It seems reasonable to conclude therefore that a higher proportion of women in building have a family background of experience in the industry than is the case for all women.

4.1.2 Educational qualifications

Academic qualifications

As expected given the nature of the sample (professional women), women with experience of the construction industry are highly qualified relative to the population as a whole, with two-fifths having a first or higher degree. An additional 15 per cent expected to gain a first degree and seven per cent a higher degree (Table 4.5).

Vocational qualifications

In terms of vocational qualifications, over half of the sample hold a HNC or HND and an additional nine per cent a national certificate or diploma or advanced GNVQ (Table 4.6). The recent

[1] If people working as managers and administrators are included, this figures rises to just over one per cent.

Table 4.6 Vocational qualifications

	Current highest vocational qualification			
	All	CIOB members	Former members	Other
First certificate or diploma/Intermediate GNVQ	2.7	2.3	1.2	5.6
N/SVQ	1.1	0.5	0.0	4.2
National certificate or diploma/Advanced GNVQ	8.8	5.4	11.1	16.9
First line supervisors	0.3	0.5	0.0	0.0
HND/C	53.1	57.5	63.0	28.2
SMETS	0.5	0.5	0.0	1.4
Other	4.0	2.7	3.7	8.5
No vocational qualification	29.5	30.8	21.0	35.2
Total cases (=100%)	373	221	81	71

Source: IES Survey 1994

nature of the introduction of NVQs is reflected in the low proportion of women currently holding these qualifications (one per cent).

A substantial minority of respondents reported no vocational qualifications (30 per cent). These women are, however, likely to hold academic or professional qualifications. A similar situation is likely to explain the proportion of women reporting no academic or no professional qualifications. There is a range of awards which meet the requirements for entry to building industry careers and there is no reason to suppose that all professionals in the industry should hold academic, vocational and professional qualifications.

Table 4.7 Professional qualifications

	Current highest professional qualification				Expected[1]
	All	CIOB members	Former members	Other	All
CIOB Associate exam	4.8	5.6	5.7	0.0	1.5
CIOB Member Exam Pt I	12.6	14.1	15.7	2.0	2.6
CIOB Member Exam Pt II	27.6	38.0	15.7	0.0	9.4
Other	15.0	15.0	11.4	20.0	4.5
No professional qualification	39.9	27.2	51.4	78.0	0.6
Total cases (= 100%)	333	213	70	50	468

Source: IES Survey 1994

[1] These data are based on the number indicating they expected to gain each qualification. It is not possible to distinguish between those who did not expect to gain the qualifications and those who did not answer the question. As a result, the column does not add to 100%.

Professional qualifications

In total, 45 per cent of respondents held a CIOB qualification as their highest professional qualification, with this proportion rising to 58 per cent among CIOB members (Table 4.7).

Where respondents indicated that they had other professional qualifications, this was usually because they were full members of one professional body but associate members of the CIOB. As a result, the proportion stating that they were CIOB members was higher (74 per cent) than that stating they had a CIOB qualification as their highest professional qualification (45 per cent).

The other main professional organisations mentioned by respondents included associate and full membership of the Royal Institution of Chartered Engineers, the Institution of Civil Engineers, the Chartered Institute of Housing and the British Institute of Architectural Technicians.

Just five per cent of respondents indicated that they had received women-only training. Of these, one fifth stated that it had led to a qualification.

4.2 Current job

4.2.1 Employment in the building industry

Over two thirds of the women surveyed are currently working in the building industry. This figure rises to 75 per cent among CIOB members but falls to 54 per cent among former members. The young age profile of the group is reflected in their experience of working in the industry: 44 per cent have worked in building for up to five years while the remainder are split almost equally between those who have been in the industry for six to nine years and 10 years or over (Figure 4.1).

The vast majority work on a full-time basis (96 per cent) (Table 4.8). Among all women in work this proportion is considerably lower (56 per cent). Even among all professional women, the group which represents a more appropriate comparison, full-time working accounts for 72 per cent of the total in employment. These figures show that part-time work among professional women in building is less common than among all women professionals. The low proportion of women in building who work part-time is almost certainly related to their relatively young age profile and their tendency not to have children. Nevertheless, even among those aged 35-44 (the age-group with the highest proportion of part-time workers), it is only just over a fifth (compared to more than two-fifths nationally).

Figure 4.1 Number of years working in the building industry

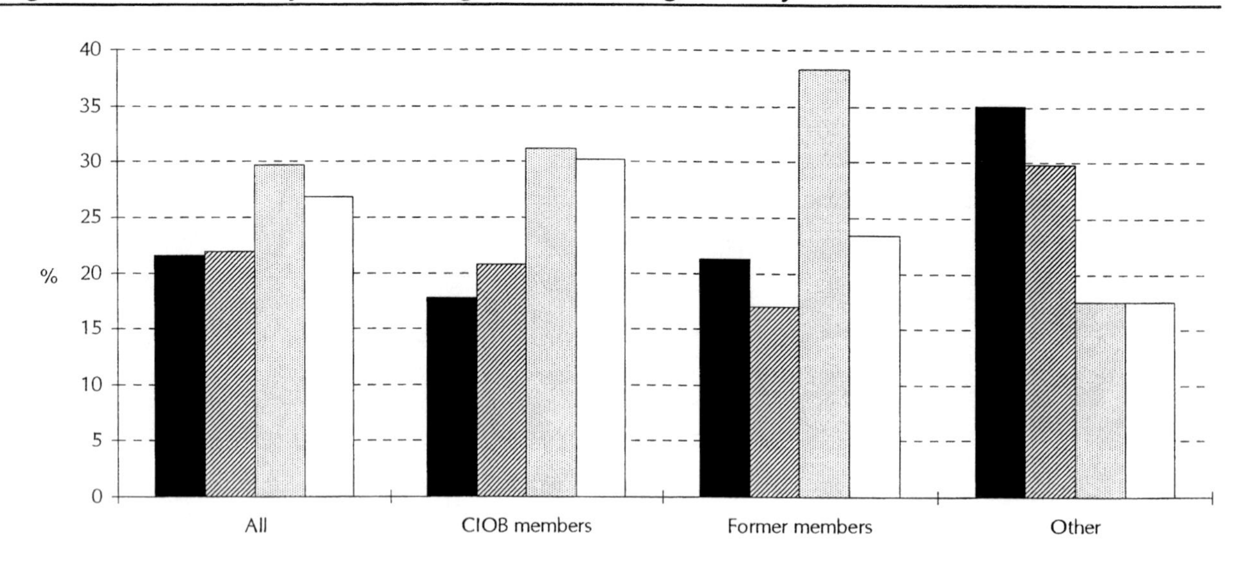

Source: IES Survey 1994

4.2.2 Type of work

The overwhelming majority of women surveyed work as employees (96 per cent), with contracting organisations accounting for the bulk of employment (60 per cent). The only other sectors to account for more than five per cent of total employment were the public sector (19 per cent) and consultancy (seven per cent).

Almost three quarters of those currently working in the building industry spend at least part of their time on-site and only 27 per cent stated that they mainly worked off-site (Figure 4.2). Thirty six per cent mainly worked on site and 37 per cent stated that their work involved an equal mix of the two types of activity. A higher proportion of CIOB members worked on-site than was the case among former members or the 'Other' group.

Table 4.8 Proportion of women currently working in building and hours of work (%)

	All	CIOB member	Former member	Other	With children
% working in building	68.8	75.0	53.7	65.6	57.9
Of which:					
Full-time (%)	95.6	94.3	98.0	98.3	81.8
Part-time (%)	4.4	5.7	2.0	1.7	18.2
Total cases	320	210	51	59	76

Source: IES Survey 1994

Figure 4.2 Main location of work

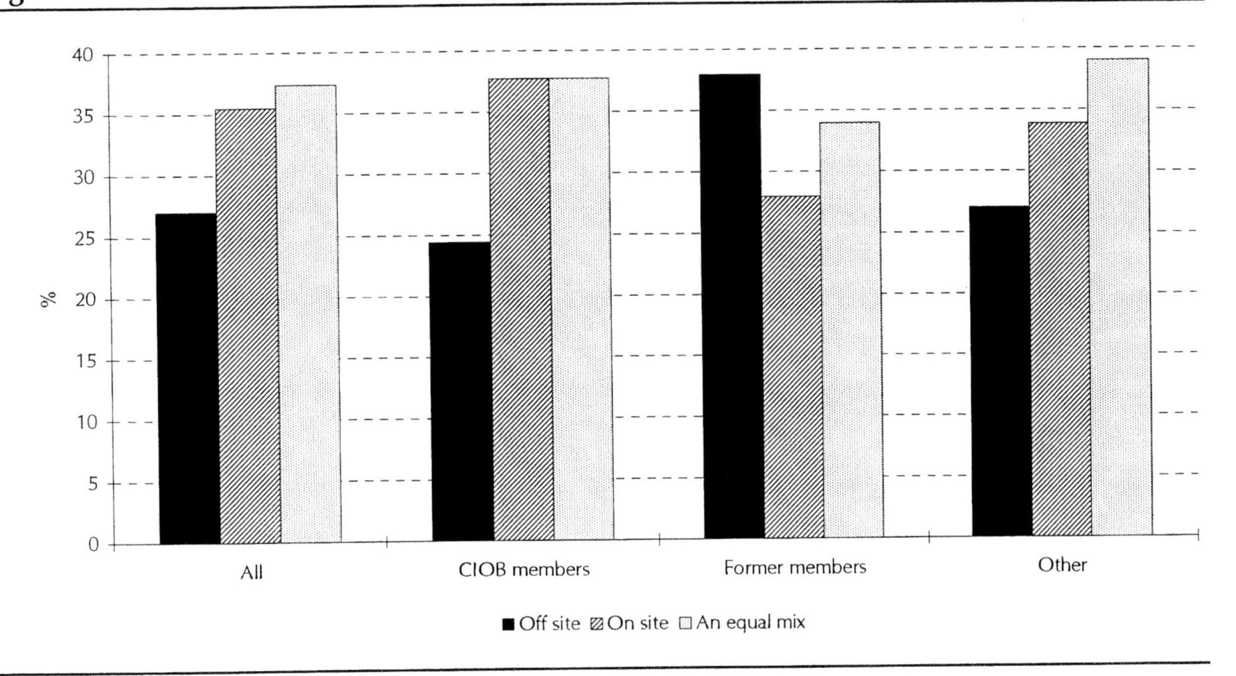

Source: IES Survey

4.2.3 Flexible working arrangements and training

In order to gain an understanding of the availability of flexible working arrangements and training activities in the workplace, we asked respondents to indicate whether a range of facilities were available in their organisation and whether they had taken them up, or would take them up if they were available.

Taking flexible working arrangements first, Table 4.9 shows that the most frequently available facilities include:

- maternity leave beyond the statutory requirements
- emergency time off
- part-time work
- flexitime.

As might be expected, in the majority of cases women working in the public sector were more likely, sometimes considerably so, to indicate that flexible working and other arrangements to help women combine a career with family life were available (see Table 4.9A in Appendix 3).

There is no national study with which to compare these figures but questions about the availability of various options were asked in a recent evaluation of Opportunity 2000 organisations (Opportunity 2000, 1994). The questions are not directly compatible with those on the Women in Building survey but they can provide some information on the kinds of facilities available in other organisations. The results suggest that building employers lag behind Opportunity 2000 employers in

Table 4.9 Availability of flexible working arrangements and training related facilities

	Available	Would take up if available	Have taken up
	All	All	All
Maternity leave (beyond statutory requirements)	44.6	41.8	4.0
Career break schemes	9.1	45.2	1.0
Job share	19.4	17.7	1.0
Part-time working	24.4	23.1	4.3
Term-time working	5.7	17.4	1.7
Flexitime	24.1	42.5	13.4
Emergency time off	56.9	28.4	15.4
Ability to work from home	16.1	40.5	7.4
After school care/holiday play scheme	5.7	30.1	1.3
Financial support for caring activities	1.0	14.0	1.0
Crèche	6.4	32.8	2.3
In house training/development (eg N/SVQs)	44.8	26.1	28.1
Professional updating or retraining	13.7	40.8	6.0
Financial support for external training	47.8	33.8	31.1
No cases	299	299	299

Source: IES Survey 1994

the facilities they provide. For example, in over 70 per cent of cases Opportunity 2000 organisations offered maternity arrangements beyond statutory requirements (compared to 45 per cent of women in the survey indicating that this was available). Part-time work was available in over 70 per cent of organisations and flexible hours in 60 per cent (in each case 24 per cent of women in building stated these facilities were available).

Other arrangements which help women combine working with family life are less widely available in building. Key among these are career break schemes (available to nine per cent of women); after school care (six per cent) and crèche facilities (six per cent). In contrast, more than two fifths of Opportunity 2000 organisations offer career breaks, although a similar proportion to those in building offer after school childcare and crèche facilities.

The arrangements favoured by most women (as indicated by whether they would take them up if available) are those which would allow them time off to have a baby (maternity leave beyond statutory requirements and career break schemes) and those which would help them combine working *full time* while raising a family (flexitime, working from home, after school childcare, crèche facilities). Part-time and term time working on the other hand are viewed somewhat less favourably. This suggests that women recognise that going part-time or working only part of the year would be more detrimental to their career than continuing in full-time work. They would, therefore, be more inclined to take up arrangements which facilitated remaining full-time work.

The greatest discrepancy between availability of an option and whether it would be taken up is in precisely those areas which would help women combine working full-time with family responsibilities. For example, 33 per cent of respondents indicated that they would use a crèche but only six per cent indicated that this option was available.

Emergency time off and flexitime are the facilities which the most women have taken up (15 and 13 per cent respectively).

Turning now to training facilities, there appears to be greater commitment to these kinds of activities than is the case for flexible or family friendly working practices. Almost a half of respondents indicated that in-house training and financial support for external training were available. In each case, around 30 per cent of women had taken up these options. Fewer had access to professional updating or retraining, although this was the option that most women indicated they would take up if it were available.

4.2.4 Partner's work

Among women working in building, 57 per cent had a partner in paid work, with the bulk of the remainder stating that they did not have a partner. Only four per cent indicated that their partner did not work.

In order to gain an assessment of the relative weight given to women's careers compared to those of their partners' we asked respondents to indicate the importance of their career when household decisions are being made. Among those whose partner was in paid work, 78 per cent indicated that equal importance was given to both partners' careers. An additional 15 per cent thought that their partner's career was accorded greater importance and seven per cent that their own career was more important (Figure 4.3).

Among women with children the proportion indicating that equal importance was given to both careers fell to 63 per cent, while the number suggesting that their partner's career was more important increased to 35 per cent.

The generally high percentage of women who stated that their careers were of equal or greater importance than those of their partner is encouraging. It suggests that managers who persist in asking about women's family lives when recruiting or

Figure 4.3 Relative importance of own partner's career by age and whether have children

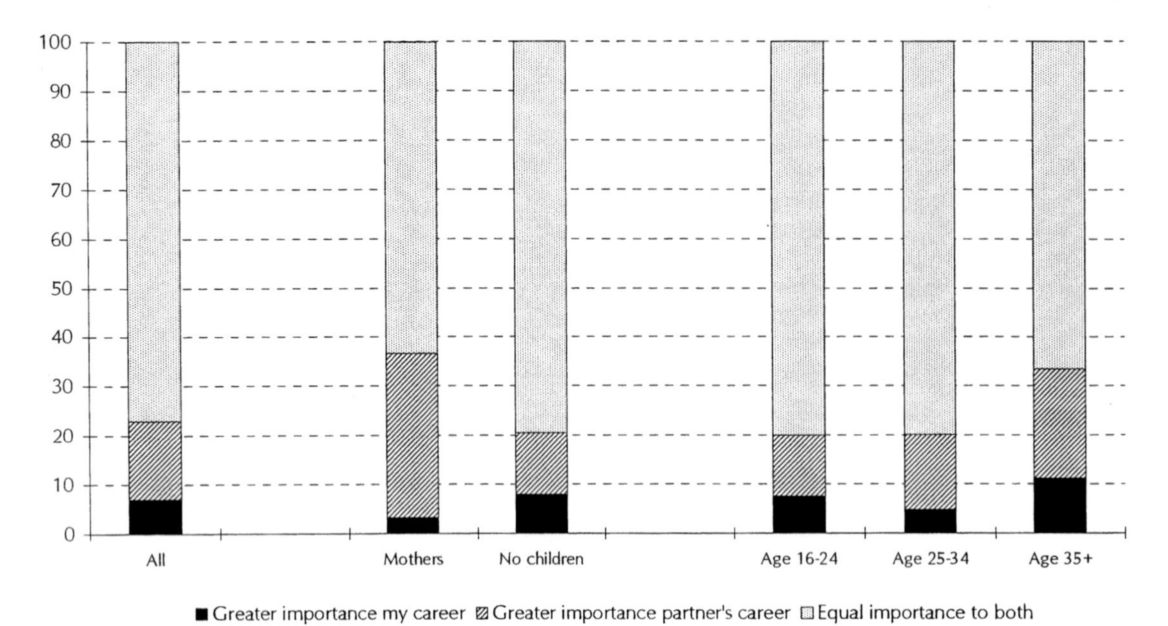

Source: IES Survey 1994

interviewing for promotion are making false assumptions about the extent to which women will allow their partner's working lives to dictate household decisions.

4.2.5 Career breaks

Only 14 women were on a career break at the time of the survey and this is too small a number for any meaningful analysis.[1] An additional 32 indicated that they had taken a break at some point in their career. Of these, more than two thirds had taken just one break, while 22 per cent had taken two breaks. The age at which respondents started their most recent break ranged from 20 to 40, with the majority starting when they were between 20 and 30 years old.

After their most recent break, 69 per cent had returned to work for the same employer. In addition, 48 per cent of those who had taken a career break were kept up to date with relevant changes in their field, although in the majority of cases (two thirds) this was paid for by the woman herself and not by her employer.

[1] Of these, when asked to describe their plans, the majority (seven out of the 12 who answered the question) expected to return to work for the same employer and only two planned to return to work but not in the building industry. The remainder of those answering the question (three) stated that they planned to return to work in the building industry but for a different employer.

Summary 4

This chapter has looked at the personal and educational characteristics and working patterns of women responding the survey. The main points to emerge include:

- women in building are a distinct group relative to all working women when it comes to their age profile (they are younger) and the proportion who are mothers (which is lower than average, even once some allowance for age differences have been made)
- they also show a marked propensity to have a family background in construction (a quarter indicated that one or both parents had worked in a technical or professional capacity within the industry)
- they are highly qualified, with over two-fifths having a first or higher degree
- the majority work full-time, even when they have children
- there is a substantial discrepancy between the current availability of various working arrangements or facilities which help women combine working with family life and those that women would take up if they were available. There is also some evidence to indicate that building employers, in particular those in the private sector, lag behind more progressive organisations in terms of the facilities they offer to women with children.
- the majority of women in building consider their careers to be of equal or greater importance than those of their partners when household decisions are being made. Managers who continue to assume the opposite are therefore basing their decisions on a poor understanding of the actual situation.

5. Career Decisions

This chapter looks at the influences on women's decisions to enter the building industry. It then moves on to consider some of the reasons they gave for leaving, or not pursuing, a career in building.

5.1 Influences on career choice

We asked respondents to assess the influence of a range of factors on their decision to pursue a career in the building industry. We used a five point scale ranging from 'strong positive influence' to 'strong negative influence'. For the purposes of this discussion, we have grouped the various factors into five main categories:

- the type of work involved
- the characteristics of building industry jobs/careers
- people outside of school
- the educational or careers advice system
- other influences.

The results for each category of influence are presented in two stages. The first looks at the extent to which each factor was influential or relevant to the career decision in the first place. These data are presented in Figure 5.1. Where respondents indicated that a factor was an influence, we then go on to examine whether the effect was positive or negative (Figure 5.2).

5.1.1 The type of work

The factors here include:

- interested in the subjects or type of work involved
- wanted to work in a job not traditionally done by women
- wanted to work in varied job locations/out-of-doors
- it's something I always wanted to do.

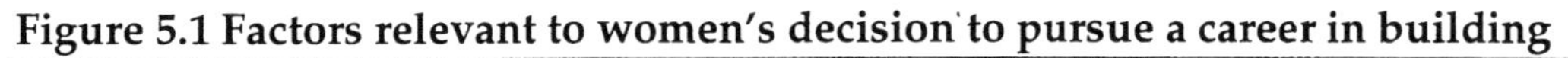

Figure 5.1 Factors relevant to women's decision to pursue a career in building

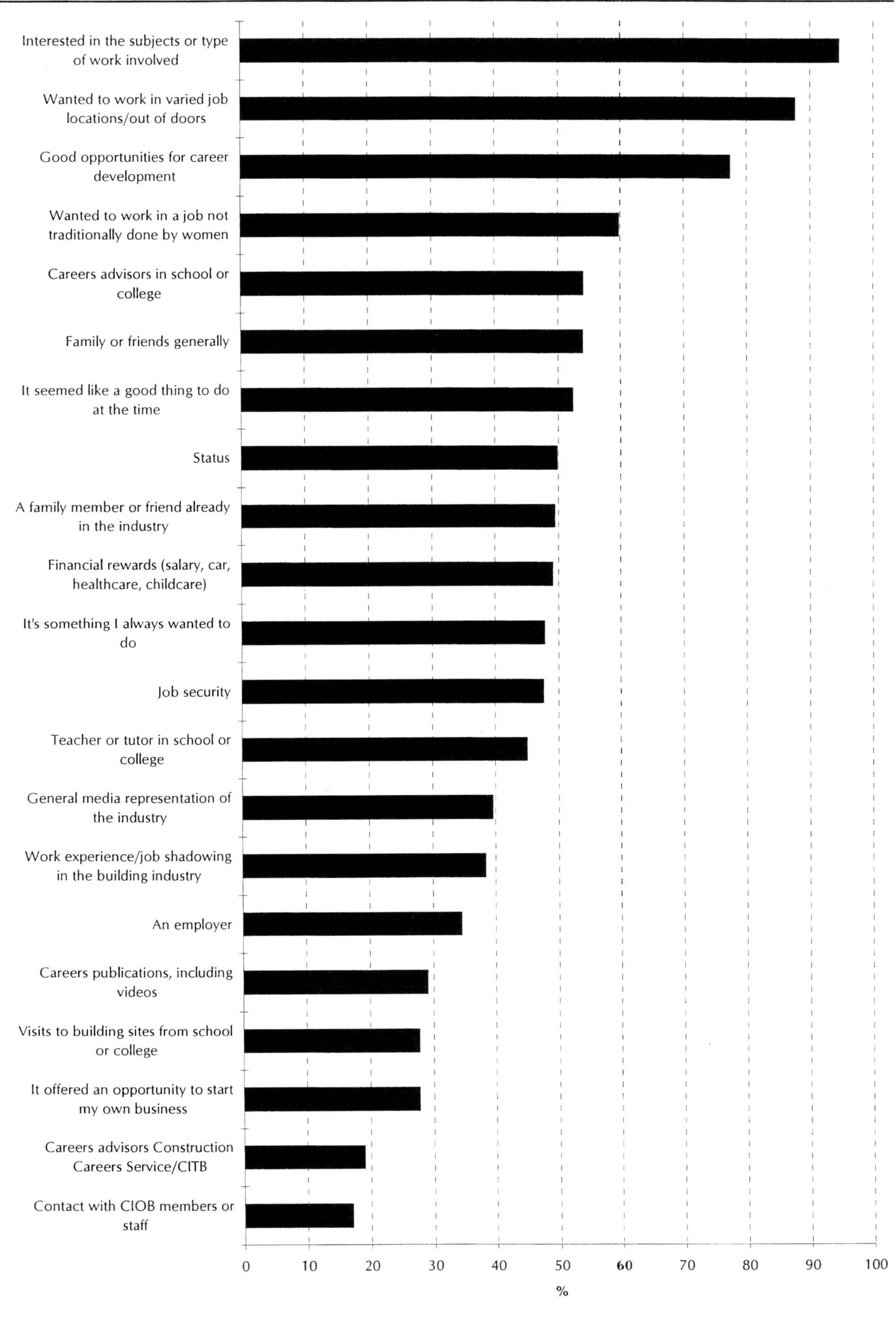

Source: IES Survey (for full text of statements see Appendix 2 Question 11)

Figure 5.2 Type of influence by factor on women's decision to pursue a career in building

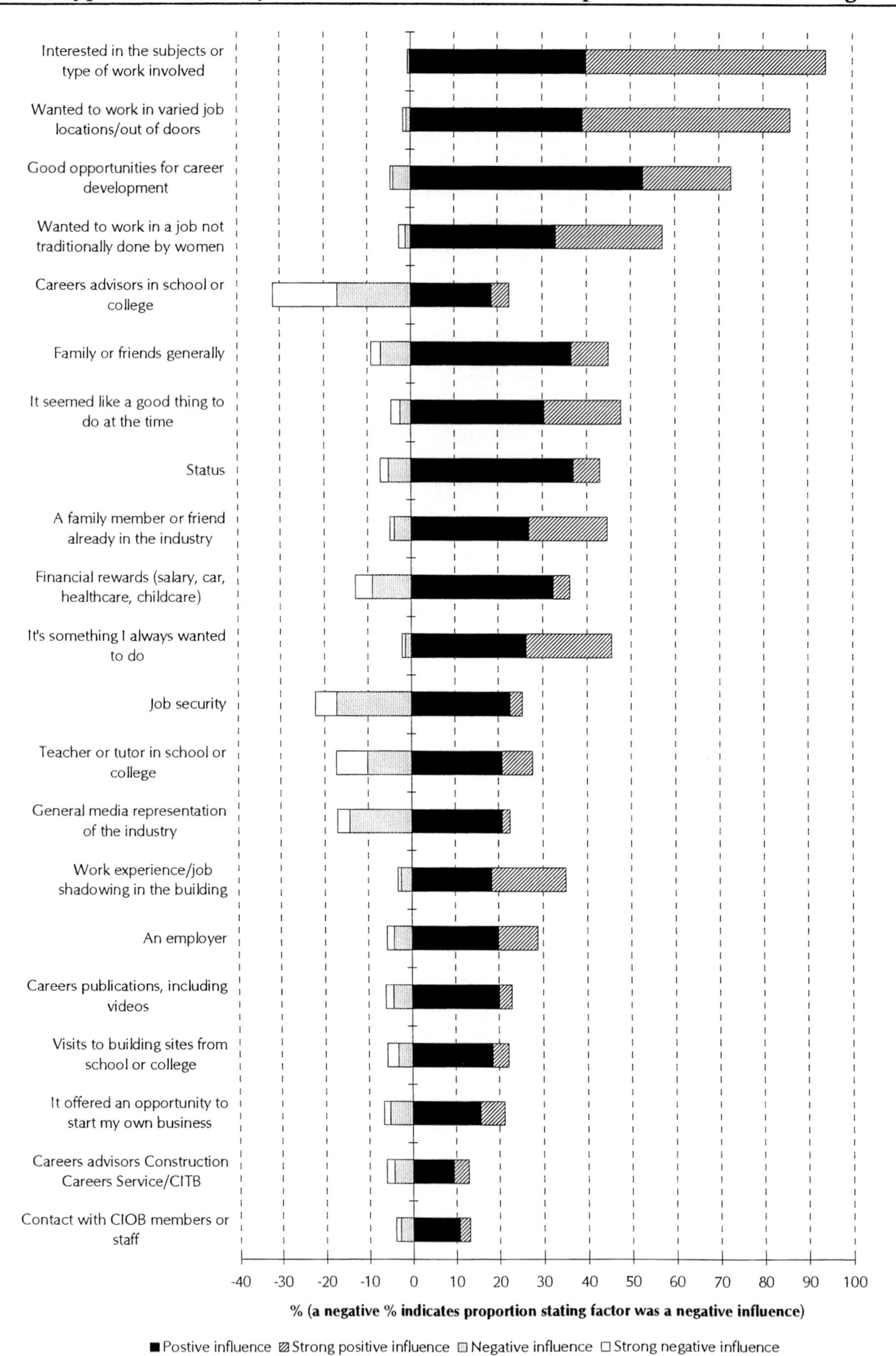

Source: IES Survey (for full text of statements see Appendix 2 Question 11)

Relevance of the type of work to career decision

This was the most clearly influential set of factors (Figure 5.1) with, in most cases, more than half of respondents stating that these issues had been relevant to their career decision. Of key significance here, with 95 per cent of women indicating that it influenced their career choice, was 'interest in the subjects or type of work involved'.

Influence of the type of work on career decision

Figure 5.2 shows the extent to which each factor positively or negatively influenced individuals' career decisions. We have taken the proportion of respondents indicating that a factor was a strong positive influence, a positive influence *etc.* and graphed the results. The bars to the left of the central line indicate the proportion of respondents indicating that the factor had been a negative influence while those to the right of the line show the proportion stating it was a positive influence. The mid-point (not an influence/not relevant) has been excluded from the presentation but it remains in the analysis — that is, on the chart, the sum of the positive and negative proportions do not add to 100 per cent with the difference being accounted for by the percentage of respondents stating that the factor was not an influence/not relevant.

The strongest positive influence in terms of the work involved was 'interest in the subjects or type of work involved'. This was followed by 'wanting to work in varied job locations/out-of-doors'. In addition, 'wanted to work in a job not traditionally done by women' was a strong positive influence for 24 per cent of respondents and a strong influence for 33 per cent.

A key point to emerge from this analysis is that the factors often highlighted as those which put women off entering technical occupations (including the fact that it is work not traditionally done by women, involves working out of doors, or requires mobility) were clearly one of the main attractions of a career in building for the women already in the industry. In seeking to attract more women into the industry, it would appear to be counterproductive to downplay these attributes. It would perhaps be more effective to highlight that work traditionally done by men can be very rewarding for women too and then seek to address the downside of working in a male dominated industry including lack of support, barriers to promotion, and problems of combining work and family life.

5.1.2 Job and career characteristics

The factors included in this category include:

- financial rewards (salary and benefits)

- job security
- status
- good opportunities for career development
- it offered an opportunity to start my own business.

Relevance of job and career characteristics to career decision

Job and career characteristics was a second set of factors which were generally quite relevant, with, 'good opportunities for career development', being of particular significance. In addition, 'financial rewards', 'job security', and 'status' were mentioned by about a half of respondents.

Influence of job and career characteristics on decision

There was less certainty regarding job status, security, financial rewards and opportunities for career development, than was the case for factors related to personal interest in the type of work involved. More women cited these as negative influences than is the case with the factors mentioned in section 5.1.1. Nevertheless, in each case the number of women stating that these were positive influences outweighed those suggesting that they were negative factors. The main exception to this was among younger women, 30 per cent of whom indicated that job security issues had been a negative influence compared to 25 per cent saying they had been positive.

These data show that while aspects of the job such as status and security were a negative influence for some, a higher proportion indicated that this was not the case. There is considerable concern in the industry about the impact of these factors on the quality of the workforce generally.

5.1.3 Influence of personal relationships

This group of factors includes:

- employers
- family member
- friends.

Relevance of personal relationships to career decision

Among factors relating to personal relationships, the one most respondents indicated had been relevant was 'family or friends generally'. This is consistent with the findings of previous studies on career decisions in construction (see Section 3.1.2). Family or friends already in the industry also influenced almost a half of respondents, while a lower proportion (34 per cent) were influenced by an employer. Contacts in the industry are, then, evidently important.

Influence of personal relationships on career decision

The strongest positive influence came from family or friends already in the industry (18 per cent). Family or friends generally were a positive influence, although here more women identified a negative effect (nine per cent compared with five per cent of those highlighting the role of family or friends already in the industry). Employers, where considered relevant to the decision, were generally seen as a positive influence, with only six per cent indicating that they exerted a negative effect.

These findings are again consistent with previous studies which have found that knowledge about working in a male dominated activity, developed via a close relationship with someone already engaged in it, can offset the negative influence generated by the general image which is portrayed. They also help explain the high proportion of women surveyed who said that their parents were in construction related technical or professional occupations (see Chapter 4).

Information derived from personal relationships about building is likely to be a key factor influencing entry. As a result, improving the existing level of knowledge about the industry is likely to be an effective method of broadening its appeal.

5.1.4 Influence of the educational and careers advice system

There were six factors in this category, including:

- careers advisers in school or college
- careers advisers from the Construction Careers Service/CITB
- teacher or tutor in school or college
- contact with CIOB members or staff
- careers publications, including videos
- visits to building sites from school or college.

Relevance of the educational and careers advice system to career decision

Here the most relevant factors were Careers Advisers in school or college (relevant for 56 per cent) and a teacher or tutor in school or college (45 per cent). Only about 20 per cent of respondents had been influenced by Careers Advisers from the Construction Careers Service or the CITB (Construction Industry Training Board) and by contact with CIOB members or staff. About 30 per cent indicated that careers publications and visits to building sites had been relevant.

Influence of the educational and careers advice system on career decision

The positive influence of the educational and careers advice system generally outweighed the negative. The exception to this is careers advisers in school or college: 32 per cent of women indicated that they had been a negative influence compared with 23 per cent receiving positive response. In addition, almost 15 per cent identified a strong negative influence from these advisers compared to only four per cent receiving a strong positive influence.

A substantial minority also identified teachers and tutors in school or college as a negative influence (17 per cent), although this was outweighed by the group who found these people a positive influence (28 per cent).

Construction industry careers advisers, where relevant, generally did a better job, with 13 per cent identifying these as a positive influence compared to the six per cent among whom the opposite was the case. Likewise, contacts with CIOB have generally been positive (in 13 per cent of cases), with only four per cent identifying this as a negative influence.

An encouraging finding about the construction industry careers advice system is that more younger people report being influenced by it than is the case among their older counterparts. In addition, for a higher proportion of the younger group, that influence was positive. Over a quarter of women aged 16-24 stated that careers advisers from the Construction Careers Service or the CITB had influenced their career choice and the vast majority of these (83 per cent) found the influence to be positive. This compares to 14 per cent of women aged 25 and over being influenced and of these less than a half finding it positive.[1] This is good evidence of the effect of recent initiatives designed to improve knowledge about the industry and women's role within it.

Careers publications and site visits had also exerted a positive influence, although only in a small minority of cases had this been strongly positive. In six per cent of cases, women had been put off by both careers publications and site visits.

---

[1] Some care in interpreting these data needs to be exercised as the number of older women influenced by construction careers advisers at all was only 36.

5.1.5 Other factors

General media representation of the industry

The general media representation of the industry was relevant to 40 per cent of respondents, with the positive influence outweighing the negative (over a fifth stated that the media had positively influenced them while 17 per cent indicated that it had been a negative influence. In neither case was the influence strong.

Work experience or job shadowing

Work experience or job shadowing was relevant to 38 per cent of respondents and was clearly a positive influence for most of them (35 per cent). Overall, 17 per cent of women indicated that work experience or job shadowing had been a strong positive influence, while only three per cent suggested that it had exerted a negative effect.

It seemed like a good thing to do at the time

This factor was included to capture the less serious side of career decisions and was, perhaps surprisingly, relevant for over a half of respondents. The majority (48 per cent) stated it had been a positive influence, with few (five per cent) indicating otherwise.

Other factors mentioned by respondents

We asked the women completing the survey to specify whether any other factors had influenced their career decisions. A total of 80 mentioned additional factors, the main ones being the positive influence of:

- that the work was challenging and satisfying (29 respondents)
- work was available/it was a job (12)
- vocational guidance (5)
- it allowed study to be combined with working (4)
- a career in building was a lifelong ambition (4)
- insight or WISE courses (3) — the small number of women mentioning this factor indicated that it had exerted a strong positive influence.

Summary 5a

The main factors relevant to respondents' decisions to enter the building industry related to the type of work involved and the characteristics of the jobs on offer. In addition, women were influenced by a range of individuals, including careers advisers in school or college, family and friends, and teachers or tutors.

The main positive influences on career choice were also the type of work involved (interest in the subjects or work involved, the attraction of working in varied job locations/out of doors and working in a job not traditionally done by women) and the characteristics of jobs (good opportunities for career development and job status).

A notable aspect of these results is the finding that the women surveyed viewed in a positive light aspects of the industry which are often seen as negative when it comes to encouraging entry by women (including working in varied locations/out of doors and in a job not traditionally done by women). This raises the issue of whether women who have selected building have different views or expectations from the majority of their peers who chose other careers. In addition, the negative influence of these factors may have been overstated.

The only factor in which the negative influence outweighed the positive was careers advisers in school or college. Since a majority of women indicated that these individuals had been relevant to their decision, this is a somewhat worrying finding. It suggests that attitudes among these advisers may need to be addressed.

The results also suggest that recent construction industry initiatives to improve careers advice are having a positive effect. The influence of construction industry advisers, while still relatively small, was greater among younger women. This influence was also more likely to have been positive than is the case for their older counterparts.

5.2 Leaving the industry

This section looks at the women who had left the building industry and their reasons for doing so.

5.2.1 Characteristics of leavers

We asked all respondents who were not currently working in the building sector or the construction industry and who were not students on building related courses about their decision to leave the industry.[1] Of the 70 women (15 per cent) who answered this question, two fifths were not working. The remainder were in paid employment, either full-time (53 per cent) or part-time (six per cent). The pattern here differed somewhat by whether or not the respondent had children. Almost all of those with children were not in paid work at all, whereas among women without children 70 per cent were in full-time and four per cent in part-time work.

Of those in work, 46 per cent had jobs as office or clerical workers, with most of the remainder being in personnel, maintenance, educational or training functions. The majority (63 per cent) did not use the skills and knowledge acquired during their training and/or work in the building industry. This proportion was particularly high among former members of CIOB, 82 per cent of whom did not use the skills they had acquired. This suggests that leaving the Institute is an indication of a change in career away from building. Given the expense of producing a skilled worker, this represents a major cost to the industry.

On a more encouraging note, 62 per cent of respondents indicated that they planned to return to work in building at some point in the future, a figure which rises to 78 per cent among current CIOB members. In addition, women with children were more likely than those without them to be planning a return to the industry (80 per cent compared to 54 per cent).[2] This suggests that one of the main reasons for leaving the industry concerns the desire to have a family, as is confirmed below.

5.2.2 Reasons for leaving

The importance of various factors in women's decisions not to remain in building are indicated in Figure 5.3. The single most important influence is the difficulty in finding work (Couldn't find work). This was extremely important for 51 per cent of those responding to this question and quite or very important for an additional 13 per cent. It was also the most significant factor for 27 per cent of leavers.

---

1 Given the nature of our sample, we were confident that those responding to the survey had some experience of building, either as students or through employment.

2 Again, these figures are based on small numbers (15 women with children and 39 without) and need to be treated with some caution.

Figure 5.3 The importance of factors in women's decision to leave building

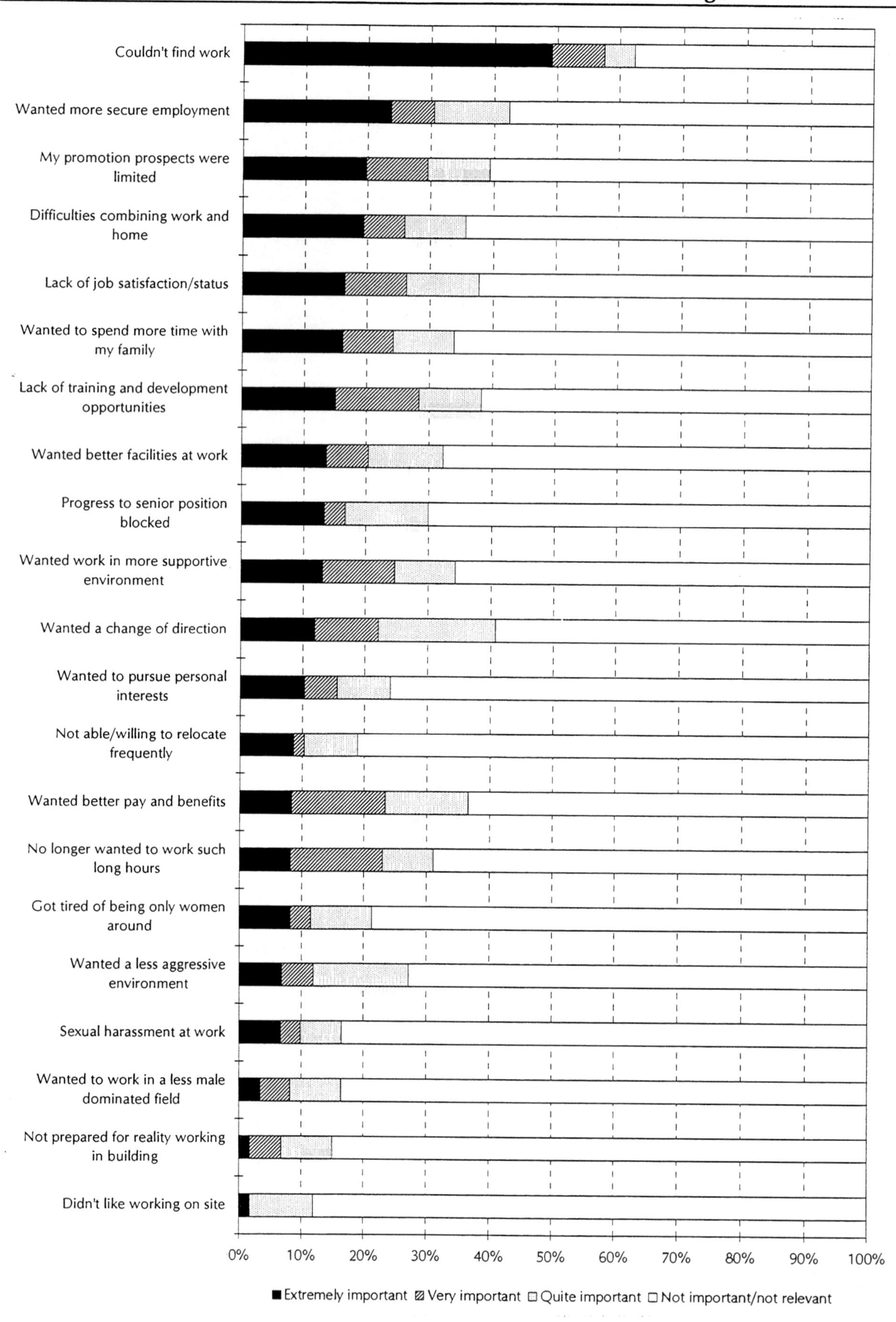

Source: IES Survey (for full text of statements see Appendix 2 Question 36)

Young people were particularly likely to identify the difficulty in finding work as extremely important in their decision to leave, with 57 per cent doing so compared to a third of older women (Figure 5.3A, Appendix 3). This reflects the particular problems facing young people as they seek to start a career in the industry.

The other major factors concerned family life, and as expected they were mainly highlighted by women with children (Figure 5.3B). Overall, 17 per cent of respondents indicated that 'wanted to spend more time with my family' was an extremely important reason for leaving building. Among women with children this proportion increased to 67 per cent. It was quite or very important for an additional 17 per cent of all respondents (and 19 per cent of women with children). For 13 per cent of the total it was the most important reason for leaving.

Likewise, the difficulty combining work and home life was extremely important for over a fifth of respondents and quite or very important for 16 per cent.

A third set of reasons concerned the characteristics of jobs in the building industry or career prospects, including:

- wanted more secure employment (identified as at least quite important by 42 per cent)
- limited promotion prospects (36 per cent)
- wanted better pay and benefits (36 per cent)
- lack of training and development opportunities (35 per cent)
- lack of job satisfaction (34 per cent)
- no longer wanted to work such long hours (31 per cent).

These are factors which in the current economic climate are as likely to affect men in the industry as women. Nevertheless, without comparable data on men, it is difficult to assess the extent to which limited promotion prospects and training/ development opportunities were more of an issue for women. The fact that 28 per cent of women indicated that one of the reasons for leaving the industry was because their progress to a senior position was blocked could be interpreted as evidence of problems moving up the career hierarchy but in the absence of more detailed information on the nature of the 'block' we cannot be sure of this.

Generally, fewer women identified more gender specific factors as important in their decision to leave the industry, although the numbers are sufficiently large to rule out complacency in this respect. One of the more worrying findings of the study is that six per cent of women leavers indicated that sexual harassment was either extremely (five per cent) or very (one per cent) important as a factor influencing their decision. An additional seven per cent identified it as quite important. Other factors which may affect women more than men include:

- wanted to work in a more supportive environment (identified as at least quite important by 31 per cent of leavers)
- wanted to work in a less aggressive environment (a quarter of leavers)
- got tired of being the only woman around (18 per cent).

The issue of wanting to work in a more supportive environment was particularly important for some women — seven per cent indicated that this was the most important reason for their decision to leave building.

It seems reasonable to suggest that many men in the industry would also enjoy greater support or less aggression at work. An improvement in these areas would therefore help to create a better working environment for all building workers while contributing to the retention of highly qualified staff (Gale, 1991a; 1992).

Few women (13 per cent) thought that they had been insufficiently prepared for the reality of working in building, and only 12 per cent indicated that 'didn't like working on site' was important in their decision not to pursue a career in the industry. This is consistent with the data on career choices described above, which showed that working in varied job locations or out-of-doors was one of the attractions of a career in building.

Summary 5b

The two main categories of people who leave the building industry are those who cannot find work and those who leave for family reasons.

The inability to find work and family or home life related concerns were the main reasons given for leaving the industry. For the majority of those answering this question, the fact that they couldn't find work was extremely important in their decision to leave the industry. Wanting to spend more time with the family and the difficulties of combining work and family life were extremely important for about a fifth of the total.

Wanting to work in a more supportive or less aggressive environment and 'got tired of being the only woman around' were also identified as important by a substantial minority of leavers.

6. Building Studies

Just over a third of respondents (159) to the survey were on a course of study related to building, the majority of which were in the 'Other' sample. We were interested in women's views on their building courses and therefore asked a series of questions about their experiences. Care will need to be taken when interpreting the data reported below as we made no attempt to ensure that the sample was representative of all women building students (see Appendix 1 for a discussion of the survey methodology).

6.1 Type of course

Two thirds of respondents who were on building related courses were engaged in part-time study. The majority of full-time courses were expected to last between three and four years. Part-time students tended to be engaged in shorter courses, with a half expected to last two years or less and an additional quarter for three years.

An encouraging factor was that over a half (57 per cent) were being sponsored by their employer, with sponsorship being particularly prevalent among students on part-time courses

A third of courses required a period of work experience, with this rising to two thirds per cent for full-time courses. This is because on full-time courses students will be less likely to be working in building while studying, requiring specific provision for work experience to be made. Almost 40 per cent of those for whom work experience is required had completed it.

6.2 Views on courses

We asked respondents to indicate the extent to which they agreed with a series of statements about their courses. A five point scale was used, ranging from strongly agree to strongly disagree. The results are presented in Figure 6.1, which shows the proportion of students who agreed or disagreed with each statement. In each case, responses from those who neither agreed nor disagreed have not been graphed. Figure 6.1A in Appendix 3, however, includes these responses.

Figure 6.1 Students' views on their courses

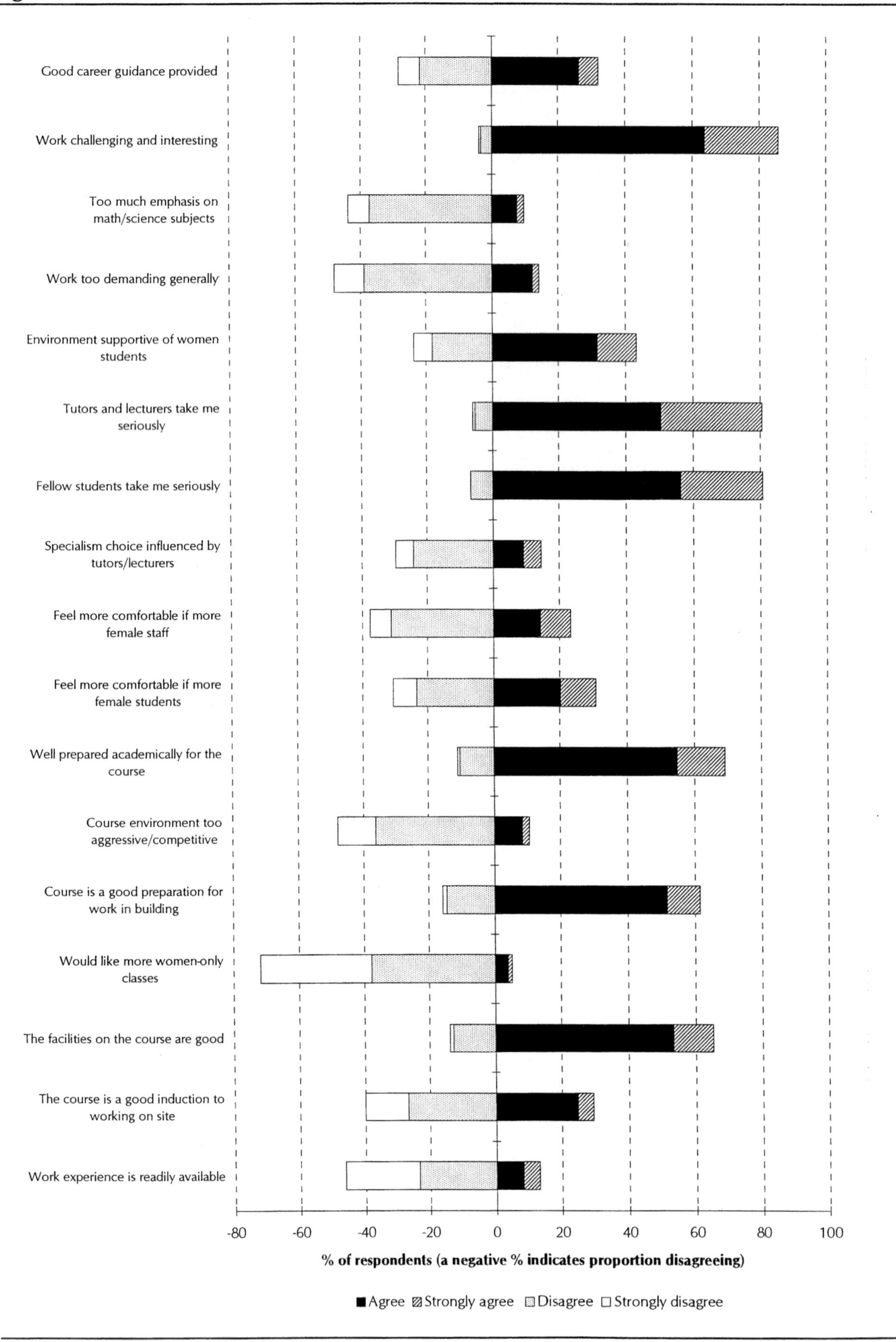

Source: IES Survey 1994 (for full text of statements see Appendix 2 Question 30)

We asked about three main areas of their courses:

- course content and academic preparation
- links between the course and a building career
- the attitudes toward, and the situation of, women on the course.

6.2.1 Course content

In accordance with the results on career choices reported in Chapter 5, 86 per cent of women thought that their course work was challenging and interesting. Less than one in ten thought that there was too much emphasis on maths and science, although a slightly higher proportion (15 per cent) considered the course was too demanding generally. Consistent with these results, almost 70 per cent of students agreed that they were well prepared academically for the course and an additional 20 per cent neither agreed nor disagreed with this statement.

Full-time and part-time students

Full-time and part-time students differed in their assessments on two points. First a higher proportion of the full-timers indicated that there was too much stress on maths and science subjects (15 per cent compared to six per cent). Second, full-time students were also more likely to agree that they were well prepared academically for the course (77 per cent compared to 64 per cent). Part-time students did not necessarily disagree with this statement, a higher proportion did, however, neither agree nor disagree (26 per cent compared to 10 per cent).

6.2.2 Link to a building career

The two main issues here are the extent to which the course was providing a good preparation for working in building and the availability of good careers guidance and work experience. Sixty one per cent of those responding to the question agreed (10 per cent strongly) that their course provided a good preparation for working in building while 16 per cent disagreed. There was, however, more criticism of the extent to which courses provided a good induction to working on site. The proportion agreeing with this statement fell to 29 per cent, while two fifths disagreed with it (including 13 per cent strongly).[1] In the absence of equivalent data for male students, it is difficult to know whether this is a general failing of courses or one more specific to women

---

[1] There were no marked (*ie* statistically significant) differences between full-time and part-time students' responses to these questions, although for both questions a higher proportion of full-time students agreed with the statements and more part-time students neither agreed nor disagreed.

than men. Nevertheless, given that in many areas of building work experience is required for progression, better preparation may help integrate women into the industry.

Turning now to the issues of careers guidance and work experience, two different patterns are evident. As regards the extent to which good careers guidance is available, slightly more students agreed than disagreed with this statement (32 per cent compared to 28 per cent). This is evidently highly dependent on the institution the student is attending, although the pattern of response to this question did not vary by full-time or part-time attendance.

There was much more agreement regarding the availability of work experience, with only 13 per cent agreeing that work experience is readily available. This evidently reflects the current economic situation of the industry but given the importance of work experience in building it is clearly an issue which needs to be addressed.

6.2.3 Situation of women

Finally, we asked a series of questions which, directly or indirectly, concerned gender. Taking the more direct statements first, we found that a substantial minority of women felt that the situation of women could be improved. There were four relevant statements here:

- the environment is supportive of women students
- I would like more women-only classes/tutorials
- I would feel more comfortable if there were more women tutors and lecturers
- I would feel more comfortable if there were more women students.

Overall, 43 per cent of women agreed that the environment was supportive of women. Nevertheless, almost a quarter disagreed, six per cent strongly, and this was particularly the case among part-time students (26 per cent disagreed with this statement compared to 21 per cent of full-time students).

There was strong opposition from both full-time and part-time students to more women-only classes. Overall, 72 per cent indicated that there should not be more women-only classes and only five per cent that there should. This is consistent with both the career choice and views on building data (Chapter 5 and 7), which show that women currently in building positively enjoy the non-traditional nature of their career and that there is general opposition to women-only activities. The response to this issue may also reflect women's realisation that they will have to learn how to cope in a male dominated environment.

As regards students and lecturers, 31 per cent would feel more comfortable if there were more women students and 23 per cent if there were more women lecturers and tutors. The responses to these questions were relatively evenly balanced, with 31 per cent disagreeing with the statement about more women students and 38 per cent disagreeing that there is a need for more women lecturers and tutors.

The more indirect gender related statements were as follows:

- the tutors and lecturers take me seriously
- fellow students take me seriously
- the course environment is too aggressive/competitive
- my choice of specialism was influenced by my tutors/ lecturers.

The vast majority of students did not indicate that they had experienced problems with being taken seriously. A small minority did, however, disagree with the first two of these statements (six and seven per cent respectively).

Few women (just over 10 per cent) thought that their course was too competitive and almost a half positively disagreed with this statement.

The final statement, included because of evidence to suggest that women more than men are channelled into particular specialisms deemed more suited to them, is difficult to interpret. Only 14 per cent indicated that they had been influenced by tutors or lecturers. In the absence of similar data on male students and more details on the type of influence exerted, however, we can not draw any conclusions about this finding.

Summary 6

These findings indicate that the majority of women on building related courses are not unhappy with their course content or the situation of women. The link between building related studies and a career in the industry, in particular regarding preparation for working on site and work experience, was of greater concern. In a sense this is not surprising because those responding to the survey will have positively chosen a non-traditional career. They are already, then, an atypical group. It is particularly encouraging to find that women students are generally taken seriously. A substantial minority, however, would like to see more women students and lecturers on their courses and feel that the environment could be more supportive of them.

cont. . . .

There were few differences in the responses of students on full-time and part-time courses. This may relate to their strongly vocational nature, which is likely to minimise the variation between different types of course.

The general level of satisfaction is indicated by the high proportion of students planning to pursue or continue with a career in the building industry (85 per cent). Only two per cent definitely stated that they were not planning to do so, with the remaining 13 per cent unsure of their future plans.

7. Views on the Industry and Strategies for Change

This chapter looks at women's views on the building industry and their assessment of the effectiveness of various actions to improve their representation and progress within it. It is divided into three main parts. The first explores the extent to which women with experience of the building industry agree or disagree with a series of statements about it. The second summarises responses to a question about how best to encourage women to enter a building career, while the final section discusses views on retention and progress.

7.1 Views on the building industry

Women respondents to the survey were asked about their views on the building industry via a series of positive and negative statements. They indicated their reaction to each statement on a five point scale ranging from strongly agree to strongly disagree (see Figure 7.1 and 7.1A in Appendix 3). The analysis is divided into four main categories:

- the work involved
- professional pay and status
- the situation of women in building
- attitudes to equal opportunities.

7.1.1 Working in building

A number of statements addressed the issue of women's views on the kind of work building professionals undertake. These included:

- The work is challenging and interesting
- The work is varied and demands a range of skills/abilities
- The work is dirty and dangerous
- Working on-site is enjoyable
- It's difficult working with people from such varied social backgrounds
- Building is a good industry for women who enjoy taking on non-traditional roles.

The pattern of response to these statements again confirmed that women working in the building professions have positively opted for the challenges involved. Respondents countered assumptions about building work which are often invoked as reasons why women are somehow inherently unsuited to it.

Taking the views on the work involved first, over 95 per cent of women responding to these questions agreed that 'the work is challenging and interesting', and that it is 'varied and demands a range of skills and abilities'.

The vast majority of women countered the view that working on-site is problematic for them, with 84 per cent agreeing with the statement that 'working on-site is enjoyable'. A third strongly agreed with this statement and only two per cent disagreed with it. By and large they did not find it difficult working with people from varied social backgrounds[1], with only eight per cent indicating that this was a problem and almost three quarters disagreeing.

A minority of women thought that the work involved was dirty and dangerous (21 per cent agreed with this statement), while two fifths disagreed. This reflects the character of the sample which focuses on women in professional roles. Nevertheless, it suggests that while the dirt and danger involved may be a factor for craft workers it is not something that the majority of professional women are overly affected by. This is a positive message for those seeking to counter the image of professional work in the industry, in particular when it comes to encouraging women to think of pursuing a career within it.

In the context of responses to these statements it is not perhaps surprising that more than two-fifths of respondents thought that building is a good industry for women who enjoy taking on non-traditional roles. This is consistent with the findings on influences on career choices discussed in Chapter 5. It does, however, suggest that in order to attract more women into the industry the advantages of non-traditional roles will have to be highlighted at an early stage.

7.1.2 Professional pay and status

There was greater ambivalence about professional pay and status. When asked to assess the statement 'The pay for professionals in building is good', almost two fifths of women indicated that they agreed while a third disagreed.

---

[1] This statement was included in order to capture women's views on working with both professionals and craft workers. Working with craft workers is sometimes assumed to create problems for women, a view which is countered by the results of this research.

Figure 7.1 Women's views on the building industry

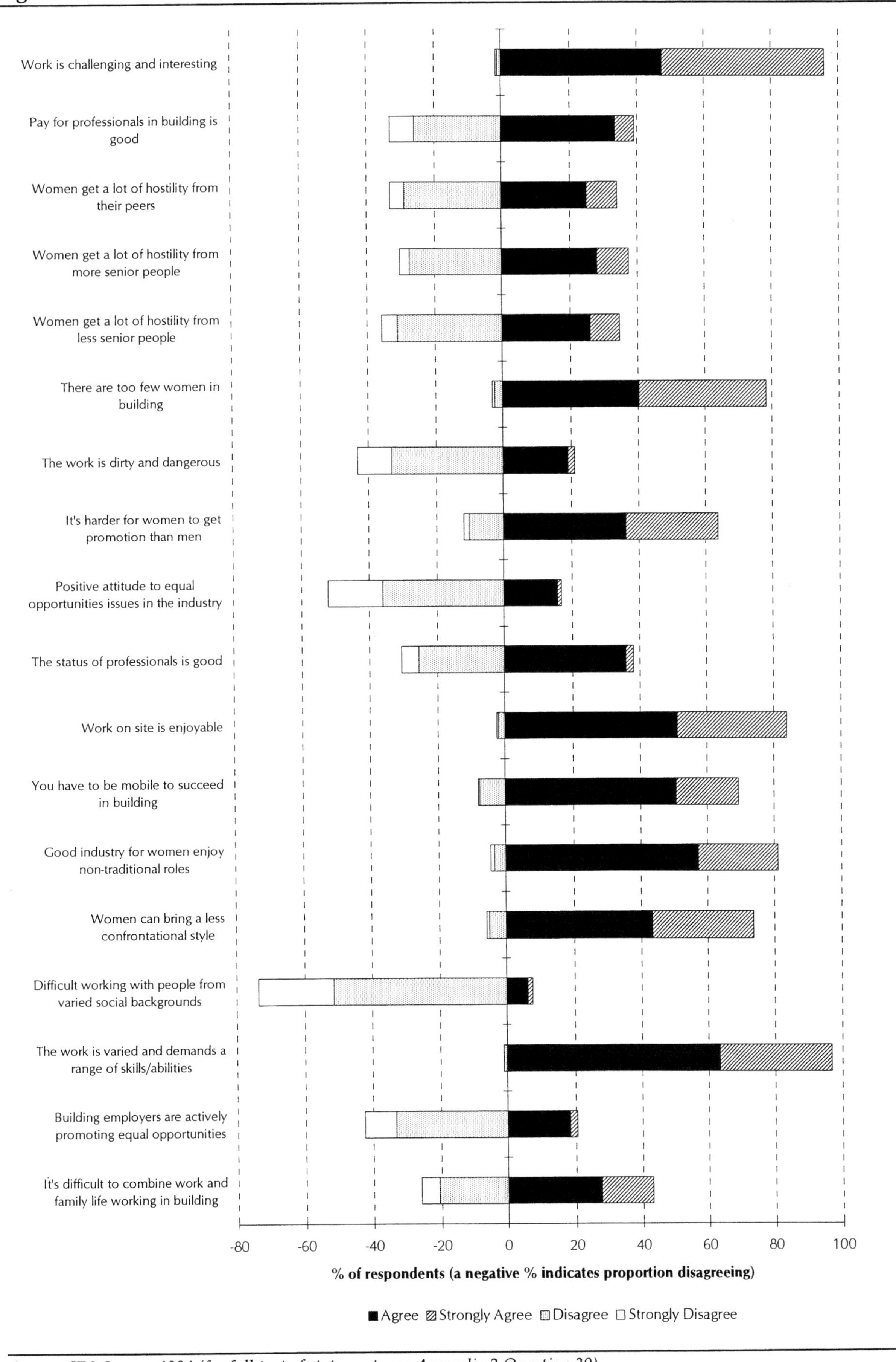

Source: IES Survey 1994 (for full text of statements see Appendix 2 Question 39)

A similar pattern was evident in the response to the statement 'The status of professionals in building is good', with 38 per cent agreeing and 31 per cent disagreeing.

Neither of these responses varied systematically by age or other personal characteristics. There was however some variation by sample group, with CIOB members being more likely to disagree than either former members or the 'Other' group (38 per cent of members disagreed with the pay statement and 34 per cent with that on status, compared to 26 per cent and about a quarter respectively of those in the other two groups). As highlighted in Chapter 4, a higher proportion of CIOB members had a professional qualification than women in either of the two other groups. They may therefore be more conscious of their professional status and more aware of the pay and status of other professions and how building compares with these. This awareness may have influenced their assessment of pay and status in the building professions.

7.1.3 The situation of women in building

In order to gain a better understanding of women's views of their working environment, we asked a series of questions about the extent to which they received hostility from co-workers (peers, more senior people, less senior people). The positive side of responses to these questions was that in each case about a third of respondents indicated that co-workers were not hostile to women. On a more negative note, however, about a third thought that there were problems in this area. This was particularly the case with more senior people, with 37 per cent of respondents agreeing with the statement 'Women get a lot of hostility from more senior people'. A slightly lower proportion (in each case, 34 per cent) agreed with similar statements referring to peers and less senior people.

Despite this, or perhaps because of it, 78 per cent agreed that there are too few women in building and only three per cent disagreed.

Two factors which are particularly likely to affect the extent to which women are enabled to remain in building professions throughout their careers concern mobility requirements and the difficulty combining work and family life. The majority (69 per cent) of women agreed that mobility is required to succeed in building. The high proportion of women with no children indicating that their own and their partner's career are of equal importance when household decisions are being made suggests that the mobility issue will become of key importance primarily when children are involved. While not all women in the industry will want to have a family, for those that do, the mobility requirement may prevent their continued career progression.

Overall, 43 per cent of women agreed that it was difficult to combine work and family life if you work in building. This proportion rose to 54 per cent for women with children. Again, not all women in the industry will be interested in having children but those that do seem likely to encounter problems. While this is the case for working women in most industries, the lack of 'family friendly' policies among building employers (see 4.2.3) suggests that for women in building the difficulties will be more marked than is the case in some other areas of activity.

7.1.4 Attitudes toward equal opportunities issues

Just over half of women disagreed with the statement 'There is a positive attitude to equal opportunities issues in the industry', 16 per cent of them strongly. This proportion considerably outweighed the 17 per cent who agreed with the statement. It is also consistent with responses to a second statement on equal opportunities 'Building employers are actively promoting equal opportunities. More than two fifths of respondents disagreed with this, compared to a fifth who agreed.

Older women also showed a more critical attitude toward the current situation. For example, while 23 per cent of younger women thought that there was a positive attitude toward equal opportunities in the industry, among those aged 25 and over this proportion fell to 11 per cent. There are two potential interpretations of this. First, that for young women the equal opportunities situation has improved. If this is the case, it would seem logical that older women would also have benefited, a situation which is not supported by the data. Second, that older women are more aware of equal opportunities issues because of their experience in the industry.

This interpretation is supported by the pattern of response to the statement 'It's harder for women to get promotion than men'. Overall, 64 per cent of respondents agreed with this and just 12 per cent disagreed. Those aged 16-24, however, were less likely to agree (52 per cent did so) than older colleagues (72 per cent).

These findings are of particular concern because of the implications they have both for women currently in building professions and those considering a career within the industry. Without a powerful commitment to equal opportunities and to changing the existing situation, the issue of improving women's entry into, and retention within, building is unlikely to be seriously addressed.

Quite apart from considerations of the building industry's need for a highly qualified workforce, the Latham Review's recent call for a change in the way the industry operates suggests that there is room for considering the effectiveness of different styles of working. Almost three quarters of respondents to the survey agreed that women can bring a less confrontational style to the industry, while only six per cent disagreed. Encouraging women into building, and providing them with the opportunity to remain and progress within it, may further the aims of reducing the level of conflict (see Gale, 1992).

7.2 Encouraging women to enter building

We have so far in this report focused on the experience and views of women working in the industry. In this and the next section we shift the focus of analysis towards the kinds of actions which could be introduced in order to improve women's representation in building professions. This part of the report looks at the potential effectiveness of various measures to encourage women to enter a career in building, while the next section is concerned with measures to enable women to remain and progress within the industry. In each case, we report the results of questions which asked respondents to assess the effectiveness of a series of suggested actions. A five point scale was used ranging from 'extremely effective' to 'would discourage women'. In addition, respondents were asked to indicate which three of the various actions proposed would be most effective.

As regards the entry of women into building, respondents were asked to evaluate the effectiveness of four main categories of options (Figure 7.2):

- actions regarding careers advice
- actions to change the image of the building industry
- actions to change the choices young women make at school
- actions designed to make the industry more attractive to women.

7.2.1 Careers advice

The main actions under this heading included:

- A campaign to change the attitudes of careers advisers
- Make information on the range of careers available
- Introduce an insight programme on building for teachers/ careers advisers
- Provide better information to parents
- Provide better information to teachers
- Highlight the good career prospects available
- Get building into the National Curriculum
- Ensure careers advice is given prior to the age of 14.

Generally, these kinds of actions were rated as either extremely or highly effective by over half of respondents. The most highly rated actions among this group were 'Make information on the range of careers available' (with 30 per cent indicating that this would be extremely effective and 42 per cent rating it very effective) and 'Highlight the good career prospects available' (with 71 per cent ranking this as at least very effective). Only slightly lower in the ranking was the introduction of an insight programme for teachers and/or careers advisers, with 69 per cent of respondents assessing this as either extremely or highly effective.

Providing better information to parents, getting building into the National Curriculum, and ensuring that careers advice was given prior to the age of 14 were ranked lower in terms of effectiveness, with just under a half of respondents indicating that these actions would be very or extremely effective. In addition, getting building into the National Curriculum was rated as 'not effective' by 16 per cent of respondents and providing better information to parents as 'not effective' by 13 per cent. With the exception of providing careers advice prior to the age of 14 and getting building into the National Curriculum[1], none of these actions were considered likely to discourage entry by women.

There was little variation by women's personal characteristics in their assessment of the effectiveness of these kinds of actions. The main difference was that a higher proportion of women with children than those without them rated changing the attitudes of careers advisers and introducing an insight programme for teachers as extremely effective. This may reflect their greater awareness of the level of knowledge careers advisers and teachers have about the industry and its effect on careers decisions.

7.2.2 Actions to change the image of building

Included in this category are:

- Counter the view of building as an unfeminine career
- Counter the dirty/dangerous image of building
- Get the media to show a more positive view.

---

[1] Even here only four per cent and two per cent respectively considered that these actions would discourage women's entry into building.

Figure 7.2 Effectiveness of actions to encourage women to enter building professions

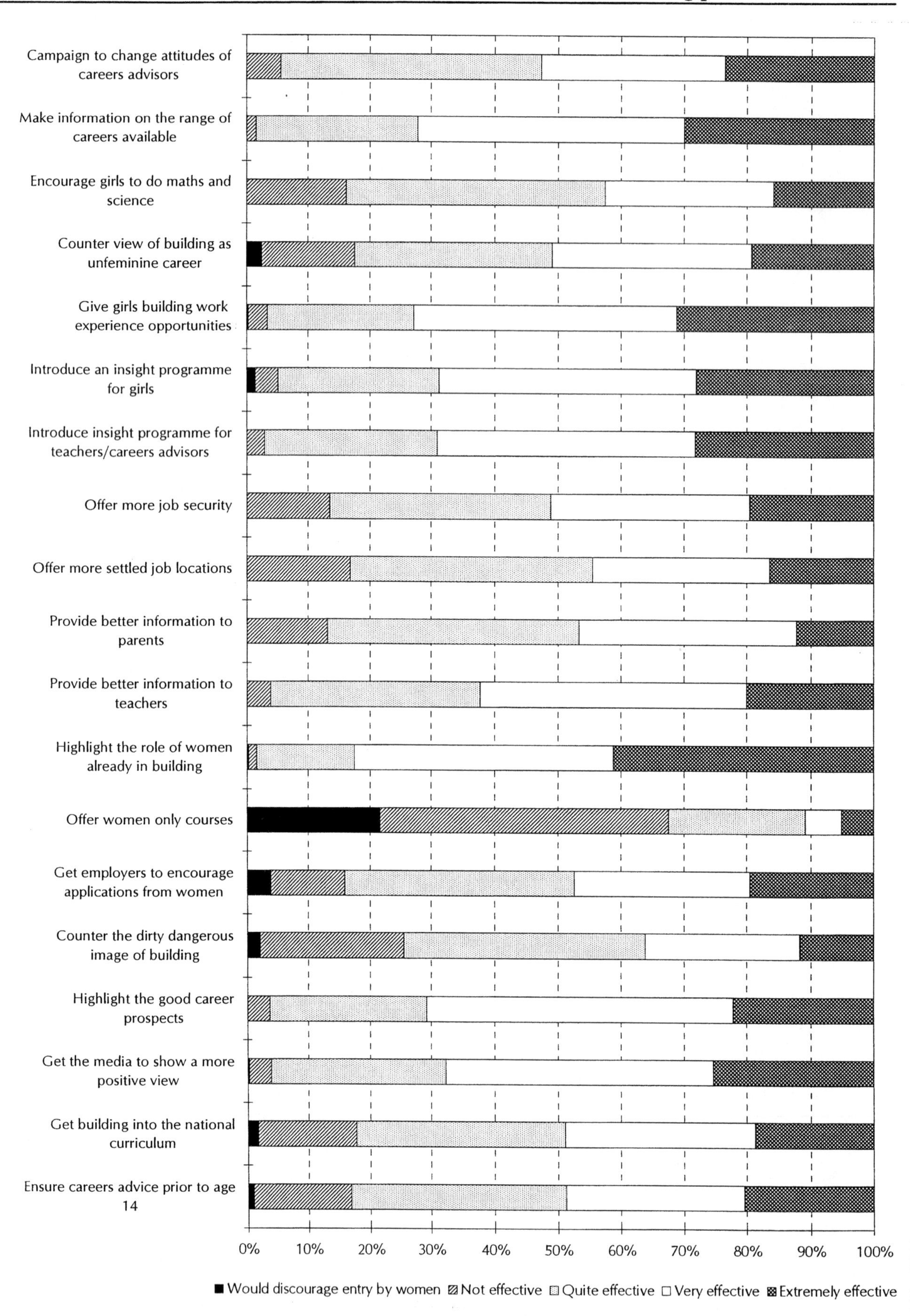

Source: IES Survey 1994 (for full text of statements see Appendix 2 Question 40)

Of these, the most highly rated action was the last, with two thirds of respondents indicating that getting the media to show a more positive view would be very or extremely effective in encouraging entry by women. Countering the view of building as an unfeminine career was looked upon favourably by just over half of respondents (51 per cent thought it would be very or extremely effective), but actions to counter the dirty and/or dangerous image of building were highly rated by only 36 per cent. This is consistent with the analysis in Section 7.1, which showed that among women already in building only a minority agreed that the work was dirty or dangerous.

Given the focus on problems with the image of the building industry in analyses of why so few women enter careers within it, the lower ranking of actions to change the image of building (with the exception of media image) may appear a surprising finding. Respondents were, however, given a range of 'careers advice' options to evaluate and their more positive assessment of these is likely to reflect awareness that the image of the industry is mediated through careers advisers, teachers and parents. Therefore, if the information available to, and advice coming from, these individuals changes (which itself may entail changing the image of building careers among this group of people), the image 'problem' will be lessened. The high rating given to encouraging a positive media image of building recognises that in addition to being influenced by individuals, young people are also profoundly affected by media images.

7.2.3 Actions to influence young women's choices at school

In contrast to the actions around careers advice, which were aimed at improving the information and advice available to young people generally (but with the specific implication that this would increase young women's knowledge and interest in building), this category focuses specifically on changing the choices young women make. The actions evaluated under this heading include:

- Encourage girls to do maths and science
- Give girls more building work experience opportunities
- Introduce an insight programme, such as site visits, for girls
- Highlight the role of women already in building
- Offer women-only courses.

This group of actions provoked the most varied range of responses. Highlighting the role of women already in building was the most highly rated action, both within this group and of all those suggested (83 per cent indicated that this would be very or extremely effective). About 70 per cent also thought that giving girls more building work experience opportunities and introducing an insight programme for girls were good ideas.

At the other end of the scale, encouraging more girls to do maths and science was perceived as at least very effective by less than half (43 per cent) of women. This ranking, however, was high relative to that given 'Offer women-only courses', with only 11 per cent indicating that this would be very or extremely effective and 22 per cent suggesting that it would discourage entry by women.

This pattern highlights a more general tension between encouraging more women to enter and remain in the industry and making women a 'special case'. There is considerable support for the former but quite marked opposition to the latter. As discussed below, this appears to relate to women's concern about being highlighted as a 'special' group and the hostility from male colleagues this may generate.

7.2.4 Actions to change jobs in building

This category of potential actions focused on the type of jobs available in building and employers activities. They included:

- Offer more job security
- Offer more settled job locations
- Get employers to encourage applications from women.

This group of actions was generally ranked lower in terms of effectiveness than many of the others. In only about half, or less, of cases was each action rated very or extremely effective and four per cent viewed encouraging applications from women as likely to discourage women from entering building.

Former members of CIOB were more likely to favour offering more job security, with 61 rating this as very or extremely effective compared to 47 per cent of current members and 56 per cent of the 'Other' group. This was raised as an issue in the discussion on decisions to leave the industry and reinforces the indications there that job security is an issue for women who have left building.

Younger women were more concerned about the effect of job security, with 23 per cent rating offering more security as highly effective compared to 17 per cent of older women. The economic environment in which young women have been introduced to the industry may account for this difference.

7.2.5 The most effective actions

Respondents were also asked to highlight the three most effective actions to improve the number of young women choosing a career in building. Taking the most effective action first, the most often mentioned options were:

- Highlight the rôle of women (16 per cent)
- Make information on the range of careers available (11 per cent)
- Give girls more work experience opportunities in building (11 per cent)
- Campaign to change the attitudes of careers advisers (nine per cent).

These options were also most often mentioned as the second and third most effective actions. Introducing an insight programme for girls and highlighting the good career prospects available were also prominent here.

7.3 Encouraging women to remain in building

This section provides details of women's assessment of the effectiveness of actions to enable or encourage women to remain in a building career and progress within it. Since almost all the options revolved in some way around equal opportunities issues, we have not divided the discussion into different categories of action as in the previous sections. It is, instead, divided into two main parts, one of which summarises the results of the survey while the second focuses more on appropriate strategies. A particular concern here is the tension between equal opportunities, on the one hand, and what is seen as positive discrimination on the other.

7.3.1 Highly effective actions

Respondents were asked to rate a range of actions in terms of their effectiveness in encouraging women to remain in the building professions (Figure 7.3). The responses show a clear preference for three kinds of actions:

- *enabling women to combine work and family life*. The most highly ranked action was introducing or improving flexible working practices and/or childcare (identified as very or extremely effective by 80 per cent of respondents). Introducing or improving career break schemes was also highly rated (with three quarters stating this would be very or highly effective).
- *commitment to, and action to ensure, equal opportunities*. Three quarters of women thought that a commitment to equal opportunities from senior managers and ensuring equal opportunities are introduced in selection, promotion, advertising and training would be very or extremely effective.
- *integrating women into the building professions* — for example by offering support to young women entering building and raising the profile of successful women, with about 70 per cent identifying these as a highly effective actions.

Figure 7.3 The effectiveness of actions to encourage women to remain in building

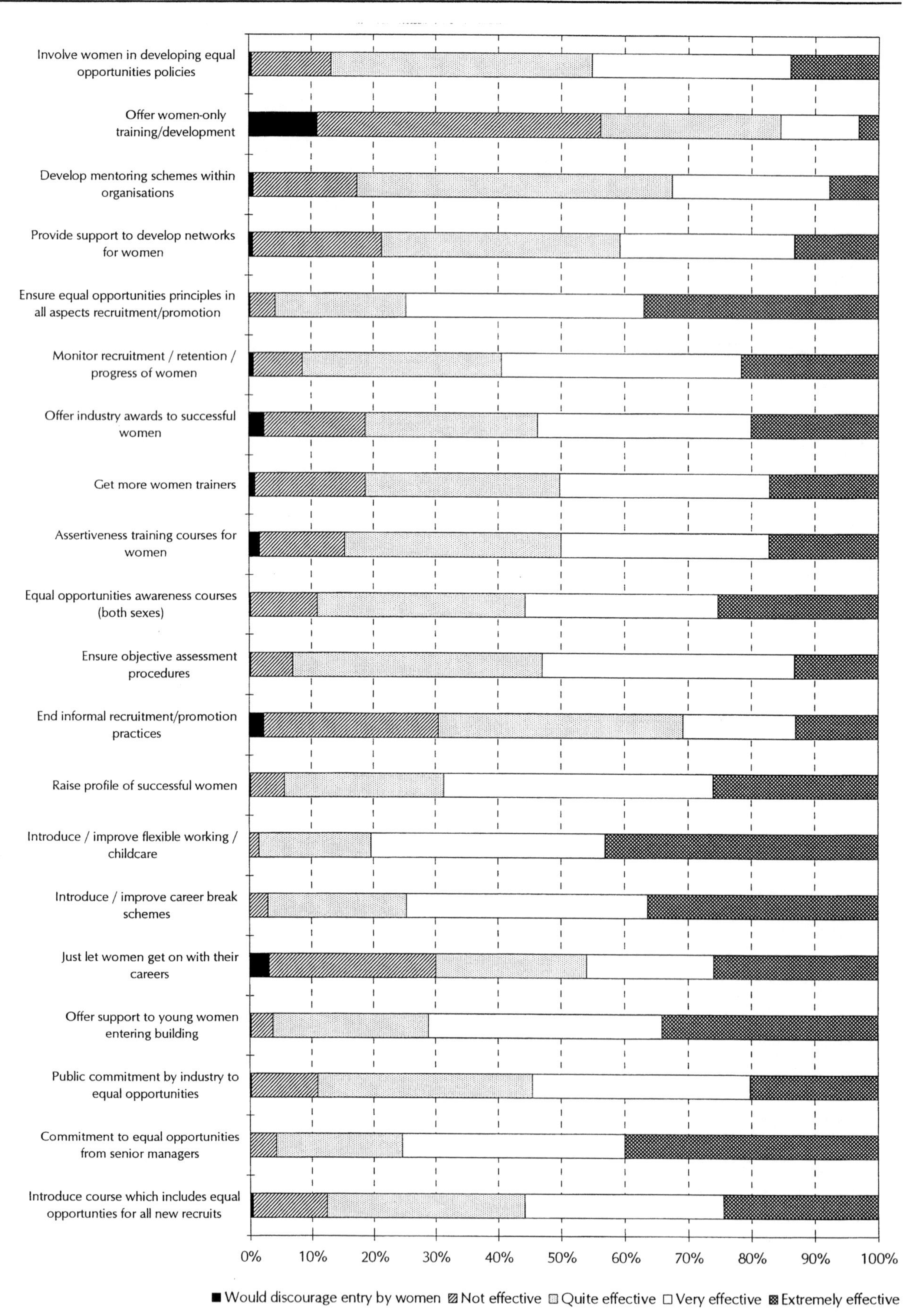

Source: IES Survey 1994 (for full text of statements see Appendix 2 Question 42)

These findings are consistent with those on the three most highly effective actions in encouraging women to remain in building careers. Of particular importance in this respect were:

- introduce or improve flexible working/childcare (16 per cent)
- ensure equal opportunities principles are applied (15 per cent)
- offer support to young people (12 per cent).

Women also made their views on their wish for equal opportunities clear in their comments on the questionnaires and in discussion groups:

> *'I just want to be treated as an equal and given the same guidance, training and opportunities as my male counterparts!' (23 year old Contract Manager)*

The situation with activities designed to help women combine work and family life is quite straightforward in that there was general support for these two options. There were, however, some quite divergent views on the general topic of equal opportunities and these are discussed in more detail below.

7.3.2 Issues around equal opportunities

As noted above, the women surveyed were quite clear about their support for equal opportunities policies and their assessment of the value of a commitment to such polices by senior management. A range of other actions, most of which related to raising the profile of equal opportunities issues were, however, viewed less favourably than a general commitment to equal opportunities. Examples include:

- public commitment to equal opportunities (identified as very or extremely effective by 55 per cent)
- induction courses which include equal opportunities awareness for all new recruits (56 per cent)
- equal opportunities awareness training courses for men and women (56 per cent).

On this latter point, the need for some kind of action was highlighted in the comments women made in the discussions and on the questionnaires:

> *'Whilst undertaking my Building Management degree course at ......, I received such remarks as "all you can do is be a secretary for a building firm". This was obviously in a joking manner, but seemed to be more the general opinion than a joke. Until the building industry removes its sexist image, I can't see why women would enjoy working in the industry, as an extremely thick skin is required.' (24 year old woman who is not currently working in building)*

'I know that I have to work twice as hard and have had to achieve twice as much to reach the present level of responsibility as male counterparts. Despite being professionally qualified, upwardly mobile, and having eight years experience, I feel as though I have reached a plateau in my career which I am finding extremely difficult to transcend. The reason for this, I was once informed by a senior colleague, is, as a single woman in her late 20s, on a one to one basis with a male peer, I would be considered a less economically viable and reliable employee given the strong possibility of my wanting to settle down and raise a family. Although I find this statement to be outrageously presumptuous and belittling of my abilities and ambitions, I do believe many employers view women in the industry in this light. And until such ill founded and biased views are overcome I hold little hope for those women in the industry, or those wishing to enter it, ever having the opportunity to experience true equality of opportunity or being judged solely on their ability to fulfil the role of a building professional.' (30 year old Building Surveyor)

In addition, there was some level of uncertainty about the effectiveness of actions to help ensure equality of opportunity:

- end informal recruitment/promotion practices (rated as very or extremely effective by 31 per cent)
- ensuring objective assessment procedures (53 per cent)
- monitoring recruitment, retention, progress of women (59 per cent).

There was also some uncertainty about actions which would highlight women as a separate group:

- involve women in developing equal opportunities policies (45 per cent rated this as very or extremely effective)
- provide support to develop network for women (41 per cent)
- get more women trainers (50 per cent)
- assertiveness courses for women (50 per cent)
- offer industry awards to successful women (54 per cent).

Of particular note in this respect was the considerable opposition to women-only training and development. Only 15 per cent of respondents thought that this would be either very or highly effective while 11 per cent indicated that it would discourage retention.

One interpretation of these patterns which is consistent with the findings of other studies is that women want equality of opportunity but tend to be more wary of policies which could be interpreted as a threat to male colleagues or which make women into a special case. Comments on the questionnaires and in discussion groups reinforced this view:

'Women have to be twice as committed to succeed in the construction industry than their male colleagues. Construction is a male dominated industry and as such trying to introduce an equal opportunities

scheme may hinder the progress of women, in as much as it may be perceived that women will be given an easy ride into the industry.' (26 year old Quantity Surveyor)

'Although we should encourage women and equal opportunities, we should not make women a 'special case' as this will only provoke resentment from the male workforce. Women must have the same opportunities within the industry, but I feel if special treatment is given, women may not get the respect they deserve.' (24 year old Construction Manager)

This means that policies designed to ensure equal opportunities have to tread a very fine line between being seen as fair to all and being perceived as positive discrimination. The latter is illegal in the UK and it is highly unlikely that employers actually practice it. There is, however, a tendency to interpret positive action (which is legal) as positive discrimination. Positive action involves activities such as encouraging women to apply for jobs or promotion and helping them to develop the necessary skills for career progression (for example, through assertiveness training courses). It does not mean that women who apply for jobs or promotion will *necessarily* get them (positive discrimination). Under equal opportunity principles, selection practices do need to be monitored to ensure fair and unbiased procedures but this is not positive discrimination. The tendency to label as positive discrimination a range of practices which seek to provide equal opportunities for women, makes it very difficult to formulate an effective and acceptable equal opportunities policy.

Summary 7

This chapter has looked at women's views on the building industry and their assessment of a range of actions designed to encourage women to enter a career in building and subsequently remain in it.

7a Views on the building industry

The clear message regarding women's views on building is that they thoroughly enjoy their jobs, including the non-traditional aspects of them.

A substantial minority (one third), however, thought that women receive a lot of hostility from co-workers (peers, senior and less senior people) and about half indicated that equal opportunities issues are not viewed in a positive light.

There was general agreement that it is difficult to combine working in building with family life.

7b Encouraging women to enter building

The most highly rated actions for encouraging women to pursue careers in the building industry were:

- actions to improve young people's knowledge about potential careers by making information on the range of careers available, and a campaign to change the attitudes of careers advisers, as well as programmes to improve teachers' and parents' information on and experience of building.
- actions to influence young women's choices, in particular by highlighting the role of women already in the industry and providing work experience opportunities.

7c Encouraging women to remain in building

Here there was wide ranging support for facilities to help women combine working with family life, and for ensuring that equal opportunities principles are introduced. Providing support for young women entering the industry and raising the profile of successful women were also highly rated.

There was, however, marked opposition to any actions which marked women out as a special group, including women-only training. The reasons for this centred around unease at provoking a negative reaction from male colleagues, and the need for women to get used to operating in a male dominated environment.

8. Improving Prospects for Women in Building

This chapter draws on the findings presented in the rest of the report, discussions with people in the industry, including the group discussions, and a range of other material to make a series of recommendations for improving the prospects of women in building. Two key categories of issues have been highlighted so far. Those concerning the small number of women who enter a career in building and the retention and progression of the women who have chosen to pursue a career in the industry. Recommendations for action in both these areas are discussed below. If these suggestions are to have any impact, however, a crucial preliminary step needs to be taken. This is establishing why the under-representation of women in building is an issue.

The chapter is divided into three main parts, each of which makes recommendations for different kinds of actions. The first relates to establishing the business case for equal opportunities in the industry. The second deals with the potential for making building a more attractive career for all, including women. The final section addresses the issue of retention and progression.[1]

8.1 Making the business case for equal opportunities

This report has documented women's experience of the building industry and the considerable barriers they need to overcome to enter and remain within it. On equality grounds it has demonstrated the need for action. Chapter 1 also discussed some of the reasons why employers are concerned about the business implications of the current situation. Employers will not, however, be influenced on equality grounds alone and many remain unconvinced by the business case presented so far.

If barriers to the full participation of women in building are to be effectively dismantled, employers need to be convinced of the arguments for taking action. Clearly, some major employers are already prepared to accept that there is an issue to be addressed. Others are not. Their major concern in this respect is indicative

---

1 In developing the recommendations outlined in this chapter, we found McRae, Devine and Lakey, 1991 and The Committee for Women in Science, Engineering and Technology, 1994 particularly helpful.

of a remarkably short-term perspective. The number of qualified men who cannot find work in the current climate is deemed to preclude the necessity of encouraging more people, be they male or female, into the industry. This perspective takes no account of future demographic and economic trends. Neither does it incorporate a vision of the need to attract the highest quality applicants, regardless of gender. From this perspective, there are sufficient people, of sufficient quality, and the business case for improving prospects for women has yet to be proved.

A few individual employers and supportive professional organisations can make a difference but more widespread change requires an industry wide initiative. The case, on business grounds, for increasing the representation of women in building needs therefore to be established amongst a wider range of employers than is currently the case. If some employers can be convinced, as they evidently are, the scope exists for making the argument to other key players.

The CIOB and other professional institutions should therefore support all existing efforts, and if necessary produce their own evidence, to demonstrate the business case for equal opportunities in the building industry. A key component of this will be publicising the reasons why some organisations are already convinced of the need to address the current gender imbalance and are taking action to do so.

8.2 Building as a career for all

This section looks at recommendations for encouraging young women to think of a career in building and addressing the needs of mature entrants to the building industry. The discussion is divided into five main parts. The first highlights the relevance of two recent reports — one concerned with the image of the building industry and the other with encouraging young women to enter careers in science and technology generally. We then move on to look in greater depth at specific recommendations on:

- initiatives aimed at schools and colleges
- mature entrants to building
- recruitment literature and activities
- other initiatives.

8.2.1 Recommendations in recent reports

The overall image of building is clearly an issue for any attempt to promote careers in the industry, as is the general topic of encouraging women to study technology and science subjects throughout the education system. Two recent reports have addressed these general concerns and made a series of recommendations for action (Latham Review Working Group 7, 1994; Committee on Women in Science, Engineering and Technology, 1994).

In terms of the industry's image, Latham Review Working Group 7 have recommended that 'a small construction industry development board be established to organise and co-ordinate all aspects of the industry's ongoing development activity', including the work of a professional promotion team (Latham Review Working Group 7, 1994, p.3).

One of the reasons young women do not choose a career in building is the general image of the industry. These proposals are clearly an opportunity to ensure that all promotion activity is monitored for its equal opportunities content and the extent to which it encourages a new view of the industry among women as well as men. A key issue here is the need to promote the range of professional careers in building and the fact that they can be done by any suitably qualified individual.

A key potential role for the CIOB in this promotional activity is to promote public awareness of women's role in the building industry.[1] We recommend therefore that the CIOB, through the WIBC, develop a strategy for promoting public awareness and encouraging media coverage of the contribution women make to the building industry. If appropriate, this should be done in co-operation with any other industry bodies established to promote the image of the industry.

Turning now to approaches aimed at encouraging more women into technology and science, the *Rising Tide* report makes two key recommendations in the field of education and training:

- Government departments for education, and education and training establishments should ensure that the initial and in-service training of teachers on equal opportunities issues includes guidance on means of maintaining the interest of girls as well as boys in all science subjects. The Office for Standards in Education (OFSTED) should routinely review the effectiveness of equal opportunities policies in schools.
- When reviewing post-GCSE courses, the relevant education departments should consider the advantages of a broader curriculum in encouraging more young people, particularly girls, to continue to study science beyond the age of 16, taking

---

[1] The *Rising Tide* report recommends that the Office of Science and Technology adopt a similar role to promote awareness of women's contribution to SET (Committee on Women in Science, Engineering and Technology, 1994, p.47).

note of the Scottish experience[1] (Committee on Women in Science, Engineering and Technology, 1994, p.35).

These recommendations are aimed at ensuring that young women are given every opportunity to study technology and science subjects. As such they are clearly relevant to the general theme of this report. Action in this area lies, however, with the relevant government education departments. While the CIOB may wish to endorse these suggestions, there are a range of more specific building related activities the industry could consider. In Chapter 3 we highlighted two key areas which need to be addressed if more women are to be attracted into careers in building:[2]

- the male dominated image (and reality) of the industry which makes young women wary of seeking a career within it, and can lead the individuals advising them to oppose building as a career option
- the lack of knowledge about the range of careers available.

The two distinct constituencies here are young women and mature entrants. While some of the initiatives discussed below are relevant to both groups, the specific needs of returners and late entrants are discussed in section 8.2.2 below. First we turn to issues of more general relevance to younger women (8.2.2).

8.2.2 Initiatives focusing on schools and colleges

There are a number of initiatives which could be adopted or extended to counter barriers to young women's entry into building careers. The key issue here is encouraging girls to think of building as an occupation relevant to them. Girls and young women do not, on the whole, wish to be seen as different or not 'normal'. The challenge facing the industry is therefore to change the view of what constitutes 'normal' interests for girls. This is an immense task and actions need to be addressed as early as possible in children's educational careers. By secondary school age many attitudes are firmly entrenched and will be that much more difficult to change.

For this reason, activities aimed at primary schools are likely to be most effective. The Curriculum Centres and similar

---

1 The report notes that the Scottish Higher Education Certificate (roughly similar to English and Welsh 'A' levels), which involves a wider range of subjects, has resulted in an increasing number of young people taking science and mathematics to higher education entrance standard over the past four years (p.33).

2 These are over and above the general image problem facing the industry which affects boys as well as girls.

educational initiatives could be of key relevance here. The former are 'local networks of schools, colleges and construction employers who jointly agree an agenda of education activities on Construction' (CITB, 1994b). The crucial advantage of these kinds of initiatives is that they raise the general awareness of the role of construction in our society and are therefore in a position to highlight its relevance to all members of that society. This has two advantages. First, they can avoid the problems associated with targeting particular groups by introducing all pupils to building, and in the process seeking to counter the idea that it is something only boys should be interested in. Second, they are in a position to influence teachers as well as pupils, thereby helping to counter the negative views on building careers some teachers have. For the purposes of encouraging women into building, these two roles can be most effectively fulfilled if the initiatives seek to raise awareness of all aspects of building and do not focus solely on the craft side of the industry.

We recommend therefore that the CIOB support the activities of the educational initiatives and seek to ensure that equal opportunities issues remain a key part of their programme. Of particular importance is the need to ensure that projects and other activities or published material are monitored for their relevance to girls as well as boys, including the language used. It is also important that women are involved in the general activities of the Centres, in particular those involving a technical aspect. In this way the message that women do pursue careers in building can be demonstrated without making their participation seem special or different.

In addition to working through the expanded network of Curriculum Centres, the building industry can take a number of additional steps to encourage women to consider careers within it. Of key importance here are activities to highlight the role of women in the industry, and to educate pupils, parents, teachers and careers advisers about the range of career opportunities available. Some specific initiatives the industry may wish to sponsor are discussed below.

General schools liaison activities

Schools liaison activities increase awareness of the opportunities in, and activities of, specific industries. A crucial issue regarding schools and colleges liaison is for employers to actively encourage women employees to participate. It is unacceptable for them to be required to take a day's leave in order to undertake these activities (as was the case with one of the women we interviewed). Liaison should be made a positive part of their working responsibilities and care should be taken to ensure that it does not simply add to the existing burden of a long working day or conflict with opportunities for career advancement. If building employers are serious about encouraging more women into building they need to support existing employees in their efforts to change the male dominated image of the industry.

We recommend that the industry encourage women's participation in general schools liaison activities and ensure that employees participating in such programmes are positively rewarded for their efforts. Liaison activities must not be pursued at the expense of the women participating in them.

The CIOB may consider a role here. It could make further use of the information generated by the 1993 survey (Lowe and Bryne, 1993) to provide information on women members in each locality and their expertise and willingness to participate in general mentoring activities (including educational initiative activities).

The CIOB may wish to consider funding some of these liaison activities. The Women's Engineering Society (WES) has a fund which provides resources for lectures and visits to schools by WES members (the Verena Holmes Lecture Fund). The CIOB's WIBC could adopt a similar strategy and help fund women builders who are willing to give up their time to participate in such activities.

A potential framework within which liaison could take place is offered by Education Business Partnerships. These co-ordinate a range of activities to help young people achieve their potential, including industry awareness for girls and the promotion of non-stereotypical attitudes and opportunities (Committee on Women in Science, Engineering and Technology, 1994, p.49).

Work placements and shadowing for girls and young women

Schools and colleges liaison can improve the general awareness of building as a career. Work placements, however, provide an opportunity genuinely to experience what a professional in building does on a daily basis. For this to be effective, the work undertaken on placements needs to be challenging and meaningful to pupils (McRae, Devine, Lakey, 1991). All pupils are now expected to complete at least two weeks' work experience before the age of 16. This is an ideal opportunity to introduce them to building.

Local employers should make every effort to encourage young women to take up work placements with them. The gender stereotype of the building industry may mean that initially this is difficult. In the longer term, however, the activities of the Curriculum Centres and other educational initiatives may ease this problem and employers need to be poised to take advantage of any indication of a change in attitudes. They can also help to foster such change themselves.

This can be done by stressing their willingness to offer work placements to local schools and colleges and suggesting that girls as well as boys may benefit from the experience. It is vital, however, that those who do take up the offer are given interesting, technical work which shows them that women are not confined to secretarial or clerical roles. Care will also need to be taken to ensure that the experience is positive and that they are not exposed to the kind of hostility some women in this study have reported.

For undergraduates on a first degree course, raising awareness among students of programmes such as STEP may be appropriate.[1] This is a work experience programme which places second year undergraduates with small and medium sized businesses. The students undertake a specific project for the company and are paid a training allowance of £100 per week. Half of this is paid by the company they are working for and half by a sponsor, which may be a private sector organisation (for example, the CIOB) or the local TEC/LEC.

As regards work shadowing, an ideal opportunity exists in the form of 'Take Your Daughters to Work', a day designated for parents to bring their daughters into work to show them what they do and give them experience of a work situation. In 1995, April 27 is scheduled as 'Take Your Daughters to Work' day. The CIOB and other industry organisations should publicise this event among building employers and encourage them and their staff (men as well as women) to participate. Site safety considerations may be a factor here but they should not be allowed automatically to rule out participation by site based employers.

Work shadowing or placements for teachers and careers advisers

Teachers and careers advisers are in a key position to influence pupils' choices. We have already highlighted the need for teacher training to encompass equal opportunities and sensitivity to how technical subjects are taught in school. We recommend in addition that building employers provide local teachers and careers advisers with work experience or work shadowing opportunities. The express purpose of this should be to encourage a new view of the industry, to develop knowledge

---

1 The Shell Technology Enterprise Programme promoted by Shell UK Limited. The national co-ordinators are STEP, 11 Bride St, London.

of the range of careers available, and to counter the view that building is an unsuitable career for a woman. Ideally, these activities should therefore be monitored for their effectiveness in these respects.

The Teacher Placement Service (TPS) organises exchange placements for teachers and people from industry.[1] The CIOB may wish to explore further current members' use and experience of this service and, if necessary and appropriate, encourage take up among teachers and building employers.

The greater involvement of local employers, TECs/LECs and education providers in local Careers Services, may provide another avenue for making careers in building for women more visible. The CIOB may wish to consider making employers more aware of the activities of TECs/LECs and Careers Services and the possibility of using these institutions to raise the level of knowledge about the range of careers in building. In this respect, it may be worth bearing in mind that Careers Services are now obliged to develop an equal opportunities policy covering client services. They may therefore be quite receptive to employers willing to encourage women into non-traditional careers.

Information on the industry

Parents, in addition to teachers and careers advisers, need to be informed of the opportunities for women in the building industry. Activities to change the existing image of building and its suitability as a career for women are clearly relevant here. Nevertheless, clear information of the kinds of careers available, which shows women doing a range of jobs, is relevant.

We recommend that the CIOB produce information for pupils, parents, teachers, lecturers and careers advisers. These will focus on working in the industry, highlighting the careers available and women's role within them. The tone and general approach of the CITB's latest book on professional careers in building has been highlighted as a good model here. They should be broadly available through schools, careers services, and libraries.

8.2.3 Initiatives to encourage mature entrants to building

The difficulty of entering a building career later on in one's life was highlighted as one of the reason's for women's under-representation in the industry by the Equal Opportunities

---

[1] 100,000 teachers have already taken placements in industry through TPS since 1989, with industry and information technology accounting for about 40 per cent of placements (Committee on Women in Science, Engineering and Technology, 1994, p.33).

Commission (EOR, 1990a). Existing education and training opportunities therefore need to be open to women of all ages, and employers should be encouraged to take on mature trainees and newly qualified employees.

Many mature people enter higher education via Access courses. The CIOB may wish to examine the potential of targeting colleges offering these courses and raising awareness of career opportunities in building among students. This is likely to be more effective if undertaken in collaboration with relevant departments in universities with Access arrangements.

We noted in Chapter 3 that women's low aspirations or confidence may affect their ability to challenge existing stereotypes. Women already in the industry who are working as secretaries or in clerical occupations may, with support and appropriate training, be encouraged to fill technical or professional roles (Fairbairn 1991, p.16). The potential of existing staff therefore also needs to be recognised and fully realised (PSI, 1994).

We recommend that the CIOB also raise awareness of existing initiatives aimed at women in which the industry can play a part. An example here is Fair Play for Women — Regional Partnerships for Equality (Employment Department Group, 1994, p52). This is an initiative which consists of regional networks of local decision makers, including employers, who plan and implement activities to increase the opportunities available to women. The CIOB may wish to monitor the extent of current involvement among building companies in Fair Play for Women and, if necessary, encourage participation in the initiative.

8.2.4 Recruitment literature and activities

Employers' recruitment literature and their participation in career talks, fairs and conventions are also a key source of information on the industry. Employers should be aware of the impact on different groups of how that information is presented, and the language used. Furthermore, recruitment literature needs to highlight the intellectual and challenging aspects of technical careers and the varied career paths available. At careers talks the participation of younger employees, especially women, will help potential applicants realise that women do play a role in building. For the demonstration effect of this to be fully realised, women's participation should focus on their professional and technical expertise and not solely their gender.

Recruitment advertising can also a play a key role in attracting applications from women. Employers and other organisations should ensure that their job or training place vacancies are displayed where women as well as men are likely to see them.

They should also contain clear reference to the recruiters' willingness to consider applications from women and ideally show women in technical roles or activities (see Women in Construction Advisory Group, 1988).

Finally, there is little point in attracting applications from women if they are then subjected to an unpleasant interview experience. This does not mean that interviewees should not be asked challenging questions, but the questions should be relevant to the job. In particular, they should not focus on personal issues. Our research has shown that the basis of the assumptions underlying many of the personal questions which are still commonly asked is wrong. It is not necessarily the case that women's careers are less valued than those of their partners' when household decisions are being made. This needs to be communicated to recruiters.

8.2.5 Other activities

There are a number of other activities and initiatives promoted by the Engineering Council which the building industry could participate in or emulate. These include:

- Engineering Council initiatives including the Neighbourhood Engineers Scheme and Opening Windows on Engineering Scheme (where local engineers give talks at schools and provide support to teachers).
- Awards for Young Engineers, for example, the Young Engineer for Britain competition.
- The Top Flight Bursary Scheme — offered by the Department for Education and The Engineering Council which give a £500 a year bursary to young people with good 'A' levels who are studying engineering.

8.3 Ensuring the retention and progression of women in building

There are three distinct sets of issues here. First, a way has to be found of supporting women in building and helping them to deal with some of the attitudes they encounter. Second, those attitudes need to be challenged. Finally, if the industry is serious about retaining expensively trained women professionals, something has to be done to help them combine working with family life. Recommendations on each of these issues are made below.

8.3.1 Supporting women in building

Our research has shown that women in building enjoy their jobs and the challenges they offer and are extremely committed to their careers. Most have developed strategies to deal with some of the downsides of working in a male dominated industry, often by becoming 'one of the lads'. This strategy of seeking to fit in with the dominant culture may be one of the reasons for the strong, and vocal, opposition to any policy which focuses on women as a 'special' group. Women fear that making them a 'special' case will create a backlash against them and undermine years of effort to become accepted and valued in their own right. In addition, they point out that women in building work in a male dominated environment and that policies which deny this reality, for example women-only courses, are likely to be ineffective.

This was often a highly contentious issue in our discussion groups. Women do not pretend to be immune to some of the attitudes they encounter (and not all have experienced problems) but most are uncertain about their ability to challenge them, or even raise them as an issue. In light of this situation, women identified having someone to discuss a difficult situation with, as being most likely to provide the kind of encouragement they needed. The situations women identified as being difficult to deal with did not necessarily relate directly to the attitudes of male colleagues — they were more likely to wish to talk with another woman about how to deal with a technical problem without making themselves vulnerable to the potential ridicule of co-workers, or provide colleagues with any cause to doubt their ability.

We recommend therefore that building employers and professional organisations encourage the establishment of local networks of women in building. The WIBC could play a key role here by simply providing women in each area with a list of other female members who have indicated that they are willing to be contacted. The aim of the network should be to provide a forum whereby women in the industry can talk with each other. It will then be up to individuals to keep the contact going. Any meetings that are convened should not necessarily focus on women's issues — there was some hostility to 'women's groups' which were strongly associated with whinging — but should encourage individuals to remain in contact and provide each other with advice and support.

One of the functions of such a network could be to provide women with a range of individuals they could contact prior to some of the informal or professional events organised by the CIOB and other professional organisations. Few of the women we spoke to regularly attended these events, because they were often the only woman there and found them extremely awkward. The network would allow women to ascertain whether any of their female colleagues were attending a particular event and arrange to accompany each other.

The discussion so far has focused on providing support for women working in building. In addition, steps could be taken to support women in building education. Increasing the number of women tutors and lecturers would help here and we recommend that every effort be made to do this. The CIOB may also wish to put student members in contact with more established women members who are prepared to undertake a mentoring role.

A second key issue concerns providing women with the skills to deal with some of the more distressing behaviour of their male colleagues (for example, being undermined in meetings and dealing with managers who refuse to speak directly to them, overt hostility) or to encourage them to be more assertive at work (for example, putting themselves forward for promotion or asking for a pay rise). Many men in the industry — especially graduates — have to deal with similar issues. Providing women and men with the opportunity to develop the skills needed to cope with these situations will not only empower them in their everyday lives, but help them deal with some of the isolation they experience. An effective course would provide women with the confidence to cope with the working environment, challenge existing attitudes and participate fully in their profession.

We recommend that courses for building professionals include an inter-personal skills training element which includes sessions on how to deal with difficult people.[1] The CIOB and large employers may also wish to consider developing such courses for existing employees. It is not necessary for these courses to focus specifically on gender issues but they do need to reflect awareness of equal opportunities issues.

Finally, some action needs to be taken to counter the negative image of women-only events. The few women we spoke to who had been on such courses found that, while they had been initially sceptical, the experience had been valuable. We recommend that the WIBC publishes a document on women-only events which highlights both the experiences of women who have benefited from them, as well as the arguments for and against such activities. This document should include information of existing courses, including non-industry specific initiatives such as SPRINGBOARD.[2]

---

[1] Several institutions already do this and their experiences could be drawn upon here (*eg* Westminster University, Reading University, Oxford Brookes University [see Wilkinson, 1993]). See also Srivastava (1992) for some of the problems women students encounter.

[2] A self development training programme to help women identify and develop their strengths and abilities (Committee on Women in Science, Engineering and Technology, 1994, p.50).

8.3.2 Changing attitudes

A substantial proportion of women we surveyed indicated that the attitude toward equal opportunities among building employers was not always positive. There were clearly some exceptions to this and they are well known in the industry. However, if building is to attract a higher proportion of well qualified women these exceptions must become the rule.

A documented equal opportunities policy is a good place to start here.[1] A potential problem here is that just as women-only policies provoked a negative reaction from many of the respondents to the survey, there was an equivalent wariness about equal opportunities. Again, there was concern about the potential of a backlash and considerable confusion about positive discrimination.[2] The reasons for adopting an equal opportunities strategy therefore need to be carefully communicated to all staff, as does its aim.

Beyond this, however, the attitude of senior and middle managers towards equal opportunities is crucial. Without demonstrated support from the top of the organisation, little in the way of change will be effected. This commitment needs to be communicated to everyone in the organisation, especially line managers. In particular, women need to know that they will be given support if they wish to challenge existing practices and behaviour, or feel that they have been unfairly treated. Only if women are supported in this way will they feel confident about making their views known.

On-site behaviour is often invoked as something it is impossible to influence. However, when contracts stipulate that certain behaviour is unacceptable, changes are made (for example, the extension to Lucy Cavendish College). There is no reason, therefore, for hostility toward or intimidation of women employees to continue. Equal opportunities policies must make this clear and be backed up by the possibility of invoking sanctions against those who refuse to change.

In addition, there is an evident need for equal opportunities principles to be introduced in recruitment and promotion

---

1. The EOC's Equality Exchange can provide further information on best practice and employers' experiences of implementing equal opportunities polices (Pollert and Rees, 1992).

2. The building industry is not alone here. The latest evaluation of Opportunity 2000 raised similar issues (Opportunity 2000). The Metropolitan Police are also having to tread a fine line between promoting equal opportunities and preventing a male backlash. The building industry may be able to learn from their experience. An approach to dealing with some of the obstacles to implementing EO policies is discussed in Pickering and Woolard (1994).

decisions. Interviewers, in particular, need to be trained in this respect and an end put to the practice of asking women about their personal lives. Open and accessible performance assessment procedures, which assess job performance against neutral and measurable criteria, would help to minimise reliance on subjective judgements about suitability. As we have seen, these are often based on erroneous assumptions.

This openness needs to extend to access to the development necessary for progress within a career. A particular issue here is mentoring. Equal access to mentoring will help women gain the support and experience needed to progress in their careers at the same rate as their male colleagues. Effective equal opportunities policies ensure that this key component of career progress is available to all who qualify on objective grounds. This may involve challenging assumptions about what women want from their careers or what work is suitable for them to do.

Finally, it is vital that equal opportunities policies are monitored and the retention and progress of all groups of employees regularly assessed. Employers should recognise equal opportunities policies as part of their overall company strategy and progress be reported in Annual Reports. Ideally, managers, including line managers, should be accountable for the implementation of equal opportunities policies. One of the most effective ways of doing this is to take an assessment of their equal opportunities record into account when making pay and promotion decisions.

The CIOB can take the lead here. We recommend that evidence of the existence, implementation and effective monitoring of an equal opportunities policy be one of the criteria for the Chartered Building Company Scheme. A organisation's commitment to equal opportunities may be measured by the extent to which it publicly supports relevant initiatives, such as Opportunity 2000.

The CIOB may also wish to consider including in its Professional Development Programme a unit on working relationships similar to that proposed in the draft of the CISC's Level 5 S/NVQ in Construction Project Management. This proposes to include as one of the elements for accreditation 'maintaining equality of opportunities in the development of working relationships'. More generally, this could be adopted by other industry organisations in their own professional updating or training programmes.

In addition to ensuring the effective implementation of equal opportunities policies, the building industry could sponsor a range of activities designed to raise awareness of equal opportunities and the profile of women in the industry. These include:

- Working with Opportunity 2000 to promote the campaign among building employers. The CIOB could also explore the potential of the 'Investors in People' initiative, which requires employers to invest in developing all their staff. Equal opportunities is an implicit, but unfortunately not explicit, part of IiP and this could be used to encourage building employers to think of all employees when making training and development decisions.
- Challenging assumptions about the efficiency of working long hours.[1] One of the main reasons why women find it so difficult to continue working in building while raising children is the prevalence of 12 hour plus days. Some of the women we spoke to pointed out that for at least part of this time their male colleagues were socialising and not actually engaged in productive work. The CIOB should seek to challenge the tendency to equate hours spent at work with hours spent working and commitment to the organisation.
- The WIBC could explore the possibility of nominating women builders for national awards, such as the *Woman of the '90s* award run by *Good Housekeeping* magazine.
- Publicising instances where common assumptions about women in the industry have been successfully challenged. For example, British Gas successfully challenged the notion that Middle Eastern clients would not accept a tender team in which a woman played a key role by asking a woman to front their proposal (Opportunity 2000, 1994).
- In order to ensure that suitably qualified women are given the opportunity to gain access to key positions within the building industry, we recommend that the CIOB keep a register of women builders qualified for appointment to boards and committees or for nomination to public appointments This information could be held centrally on a CIOB database and updated regularly.
- The CIOB may also wish to compile a register of Continuing Professional Development (CPD) and management training courses and disseminate this information widely, especially to women on career breaks. This would include women-only and assertiveness training courses and seek to ensure that information on the range of training opportunities available is brought to the attention of all employees.

A final point here relates to the CIOB itself. It will be quite simple for the Institute to stop sending out letters and other literature which assume a male audience and to take the lead in recognising in all its activities that women are an increasing proportion of the membership. We recommend that the target

---

1 This is particularly the case now that employees can sue for compensation for stress related illness.

audience should be assumed to comprise both men and women and all the Institute's output worded accordingly.

8.3.3 Enabling women to combine work and family life

Enabling women to combine work in building with family life is one, if not the, key factor in encouraging them to remain in the industry. Not all women want children. Many do, however, and at present they find it very difficult to combine both roles.

Of prime importance here is the provision of affordable and accessible childcare. We recommend that the CIOB take a number of steps on this issue:

- Publicise its support for increased provision of publicly funded and locally available childcare services, including out-of-school schemes.
- Encourage building employers to join organisations campaigning for publicly funded childcare (*eg* Employers for Childcare).
- Join other organisations in lobbying for childcare costs to be claimable against employees' tax (in short, tax relief on childcare).
- Ensure that building employers are aware that companies can offset the cost of childcare facilities against corporation tax and that these facilities can be used tax-free by employees.
- Encourage employers to make use of the childcare voucher system for all parents.
- Encourage the provision of childcare facilities by employers.

Other practical steps to help women combine work and family responsibilities include the provision of career breaks, maintaining contact during a break from work, facilitating the return to work (for example, through re-entry training or part-time work/job-sharing) and providing a range of family friendly policies.

There is tremendous opposition on cost and 'labour surplus' grounds to these kinds of actions. We recommend therefore that the CIOB collects and disseminates information on the economic and other benefits of family friendly policies within the industry and/or in other, preferably related, industries. This could be done as part of the development of the business case for equal opportunities (recommended above).

One facility which could be offered on a broader basis relatively easily is to make working from home available to parents. This may be difficult for those with on-site responsibilities but many of the tasks building professionals undertake can be done from an appropriately equipped home.

A long term recommendation is that a range of family-friendly policies be introduced by all building employers and be available to all employees.

Bibliography

Anderton F, (1986) 'Confessions From a Building Site', *Architects Journal*, 10 September, Vol. 184, No. 37

Barr B, (1994), 'Time to hunt down this beast of on-site abuse', *Construction News*, August 18, p 12-13

Beacock P M, Pearson J S D and Massey H P, (1989) *Characteristics of Higher Education for the Construction Professions Development Services*, Project Report 24, CNAA, London, pp 96-7

Bale J, (1985) 'Building for a New Age', *Building*, 24 May, Vol. 248, No. 21, pp 38-39

Billingham E, (1994) 'Graduate recruitment numbers perk up', *New Builder*, 26 August 1994, p 7

Bolton A, (1994) 'Industry study reveals endemic discrimination', *New Builder*, 5/12, August 1994, pp 6-7

Boulgarides J D, (1985) 'Decision Styles, Values and Characteristics of Women Architects In the United States', *Equal Opportunities International*, Vol. 4, No. 3, pp 1-11

Brannen J, Meszaros G, Moss P and Poland G, (1994) *Employment and Family Life: A review of research in the UK, (1980-1994)*, Employment Department Research Series No. 41, November

Breakwell G M and Weinburger B, (1983) *The Right women for the Job: Recruiting Women Engineering Trainees*, Report of the Training Division of the Manpower Services Commission, Sheffield

Bryne E, (1992) 'Role Modelling Out — Mentorship In!' *Woman Engineer*, Vol. 14, No. 16, Spring, pp 4-8

Callender C, J Toye and H Connor, (1992) *Evaluation of the Technician Engineer Scholarship Scheme (TESS)*, IMS, Brighton

Carter C and Kirkup G, (1990) *Women in Engineering: A Good Place to Be?* Macmillan

CDP (Committee of Directors of Polytechnics), (1988) *First Destinations of Polytechnic Students Qualifying in 1988*, CDP, London

Chevin D, (1994) 'Ice breaker', *Building*, 2 December, pp 28

Chevin D, (1995) 'Sex Appeal', *Building*, 13 January, pp 18-21

Chisholm L A and Holland J, (1986) 'Girls and Occupational Choice: Anti-sexism in Action in a Curriculum Development Project', *British Journal of Sociology of Education*, Vol. 7, No. 4, pp 353-365

Chivers G, (1986) 'Intervention Strategies to Increase the Proportion of Girls and Women Studying and Pursuing careers in Technological Fields: A Western European Overview', *European Journal of Engineering Education*, Vol. 11, No. 3, pp 247-155

CIOB (1989) *Building Education for Tomorrow*, Education Strategy Working Party

CITB (1994a), *The Construction Industry Handbook 1993 and 1994*, CITB

CITB (1994b), *Prevocational Education: Curriculum Centre Statistical Analysis 1993-1994*, CITB

Clarke K, (1991) *Women and training: a review*, Equal Opportunities Commission Research Discussion Series 1, EOC, Manchester

Cockburn C, (1985) *Machinery of Dominance: Women, Men and Technical Knowhow*, Guernsey Press Company

Cockburn C, (1987) *Two Track Training: Sex Inequalities in the Youth Training Scheme*, Macmillan, London

Cockburn C, (1991) *In the Way of Women: Men's Resistance to Sex Equality in Organisations*, Basingstoke, Macmillan

Committee on Women in Science, Engineering and Technology (1994) *The Rising Tide: A report on women in science, engineering and technology*, HMSO, London

Corcoran-Nantes, Y and K Roberts, (1995) '"We've got one of those": The peripheral status of women in male dominated industries', *Gender, Work and Organisation*, 2:1, pp 21-33

Court G, (1995) *Women in the labour market: trends and issues over the past 20 years*, Institute for Employment Studies, draft report, January

Court G and Meager N, (1994) *Higher Education in the UK*, Paper prepared for 2nd meeting of International Higher Education Research Network, WZB, Berlin, December

Courtenay G and McAleese I, (1994) *England and Wales Youth Cohort Study: Cohort 4: Young people 17-18 years old in 1990 — Report on Sweep 2*, Employment Department Research Series Youth Cohort Report No. 27, March

Crompton R and Sanderson K, (1990) *Gendered Jobs and Social Change*, London, Unwin and Hyman

Crompton R, (1992) 'Where did all the bright girls go? Women's higher education and employment since 1964', in N Abercrombie and A Warde (eds), *Social Change in Britain*, Polity Press, Cambridge, pp 54-69

Crompton R, (1994) 'Occupational Trends and Women's Employment Patterns', in R Lindley (ed.), *Labour Market Structures and Prospects for Women*, Equal Opportunities Commission, Manchester

Crompton R, Hantrais L and Walters P, (1990) 'Gender relations and employment', *British Journal of Sociology*, 41:3, pp 329-350

CSU (Central Services Unit) (1992) *First Destinations of Students from the New Universities Qualifying in 1992*, CSU, Manchester

Dainty, A (1993), *Maid for the Job: Attracting women to the construction industry*, BSc Building Final Year Dissertation, The University of Glamorgan, Department of Civil Engineering and Building, April 1993 (2 volumes)

Davies C, (1991) 'Part-Time Possibilities', *Architects Journal*, 20 March, pp 40-41

Davies J, (1987) 'Demolishing', *Building*, 8 May, Vol. 252, No. 7495 (19) pp 38

Davies M E, (1993) *Women in Construction Attracting Women into Construction Industry*, dissertation to University of Central England in Birmingham in partial fulfilment of the requirements for the award of BSc Hons, April

Davis L, (1994) 'First Among Equals', *Building Economist*, January, pp 16

Dean Y and Francis S, (1991) 'A Way in For Women', *Architects Journal*, 20 March, pp 43-45

Deem R (ed.), (1984) *Co-education Reconsidered*, Open University Press, Milton Keynes

Delamont S, (1989) *Knowledgeable Women: Structuralism and the reproduction of elites*, Routledge, London

Devine F, (1991) 'Women in Engineering and Science: Employers Policies and Practices', *Woman Engineer*, Summer, pp 7-9

Devine F, (1992a) 'Gender Segregation in the Engineering and Science Professions: a case of continuity and change', *Work, Employment and Society*, 6:4, pp 557-575

Devine F, (1992b) 'Gender Segregation and Labour Supply: On Choosing Gender Atypical Jobs', *British Journal of Education and Work*, Vol. 6, No. 3, pp 61-74

DfE Statistical Bulletin, (1993) *Statistics of further education college students in England 1970/71-1991/92, 14/93*, Department for Education, June

DfE Statistical Bulletin, (1994a) *GCSE and GCE A/AS Examination Results 1992/93, 7/94*, Department for Education, June

DfE Statistical Bulletin, (1994b) *Student Numbers in Higher Education — Great Britain 1982/83 to 1992/93, 13/94*, Department for Education, August

DfE, (1994c) *Education Statistics for the United Kingdom 1993*, HMSO, London

DfE, (1994d) *Higher Education Statistics*, HMSO, London

Edwards C, (1995), 'Radical restructuring leaves women out of management', *People Management*, 12 January, p 55

EITB, (1983) *Insight: A Review of the Insight Programme to Encourage More Girls to Become Professional Engineers*, EITB, Occasional Paper No. 10

EITB, (1987a) *Insight: Encouraging girls to become professional engineers*, EITB, RC18

EITB, (1987b) *Women in engineering: EITB initiatives*, EITB, RC19

Elison R, (1994) 'British labour force projections: 1994 to 2006', *Employment Gazette*, April, pp 111-121

Employment Department Group, (1994) *United Nations Fourth World Conference on Women — Beijing 1995: Report of the United Kingdom of Great Britain and Northern Ireland*

Employment Gazette, (1994) 'Labour Force Survey Helpline', December

Engineering Council, (1987) *Science and Engineering Degree Courses Women Entrants 1982/3-1986/7*, London

EOR, (1990a) 'Women in Construction', *Equal Opportunities Review* No. 30, March/April, pp 16-22

EOR, (1990b) 'Women in Engineering', *Equal Opportunities Review* No. 34, Nov/Dec, pp 11-17

Evetts J, (1994) 'Women and Career in Engineering: Continuity and Change in the Organisation', *Work, Employment and Society*, Vol. 8, No. 1, pp 101-112

Evetts J, (1990) *Women and Career: Themes and Issues in Advanced Industrial Societies*, London, Longman

Fairbairn D, (1991), *Profiting from change in the recruitment of women: opening doors for women without dropping bricks*, paper presented to Construction Industry Training Officers' Conference, Profiting from Change, Coventry, 17-19 April

Fuller A, (1991) 'There's more to science and skills shortages than demography and economics: attitudes to science and technology degrees and careers', *Studies in Higher Education*, 16:3, pp 333-341

Furlong A, (1986) 'Schools and the Structure of Female Occupational Aspirations', *British Journal of Sociology of Education*, Vol. 7, No. 4, pp 367-377

Gale A W, (1987) 'The Socio-Economic Significance of an Increase in the Proportion of Women in the Construction Industry', *Managing Construction Education*, Coombe Lodge Report, FESC, Vol. 20, No. 2, pp 113-124

Gale A W, (1989a) 'Women in the British Construction Professions: A discussion of pilot research', in *Proceedings of Gender and Science and Technology 5th International Conference, Vol. 2: In college and at work*, September 17-22, Haifa, Israel, pp 134-431

Gale A, (1989b) 'Attracting Women to Construction', *Chartered Builder*, September/October, pp 8-13

Gale A W, (1991a) 'What is Good for Women is Good for Men: Action Research Aimed at Increasing the Proportion of Women in Construction', in Barret P and Males R, *Practice Management: New Perspectives for the Construction Profession*, Chapman and Hall, pp 26-34

Gale A W, (1991b) *Construction Industry Careers Experience Course 8-10 April 1991 at the University of Reading: All female participants, First Interim Evaluation Report*, Department of Building Engineering, UMIST

Gale A W, (1991c) *Construction Industry Careers Experience Course 16-18 July at the University of Salford: Male and female participants in single sex groups, Second Interim Evaluation Report*, Department of Building Engineering, UMIST

Gale A W, (1992) 'The construction industry's male culture must feminize if conflict is to be reduced: the role of education as a gatekeeper to a male construction industry', in P Fenn and R

Gameson (eds), *Construction Conflict Management and Resolution*, E & F N Spon, pp 416-427

GHN (1995) *Future Top Managers*, GHN Career Management Consultants, London

GHS (1992), *General Household Survey* 1992, HMSO, London

Gibbins C, (1994) Women and training — data from the Labour Force Survey, *Employment Gazette*, November, pp 391-401

Greed C, (1990a) 'Women: has RICS failed?', *Chartered Surveyor Weekly*, February 22, pp 108

Greed C, (1990b) 'How to Draw Women in', *Chartered Surveyor Weekly*, 1 March, pp 140

Greed C, (1990c) 'Surveying practice: Is there a problem?' *Chartered Surveyor Weekly*, 8 March, pp 142

Greed C, (1991) *Surveying sisters: women in a traditional male profession*, Routledge, London

Greed C H, (1994) *Women and planning: creating gendered realities*, Routledge, London

Hacker S, (1989) *Pleasure, Power and Technology: Some Tales of Gender Engineering and the Cooperatives Workplace*, Unwin Hyman, London

Hacker S L, Smith D E and Turner S M (eds), (1990) *Doing It the Hard Way: Investigations of Gender and Technology*, Unwin Hyman

Hansard Society Commission, (1990) *Women at the Top*, The Report of the Hansard Society Commission

Harris (The Harris Research Centre), (1988a) *Factors Affecting Recruitment for the Construction Industry: The view of young people and their parents*, CITB, King's Lynn

Harris (The Harris Research Centre), (1988b) *Factors Affecting Recruitment for the Construction Industry: The view of employers*, CITB, King's Lynn

Harris (The Harris Research Centre), (1989a) *Report on survey of careers advisors*, CITB

Harris (The Harris Research Centre), (1989) *Report on survey of undergraduates and sixth formers*, CITB, King's Lynn

Hermon A (1994), 'Women face gauntlet of harassment', *Western Morning News*, 3 September

Hirsh W and Jackson C, (1990) *Women into Management: Issues influencing the entry of women into managerial jobs*, IMS Paper No. 158, Brighton

Hodkinson, (1994) 'How Young People Make Career Decisions', *Skills Focus*, Issue 6, Autumn, pp 1-3

Hood C, Peacock S and Smith J, (1984) *Girls and Technical Engineering: A Report of the Pilot Programme to Encourage More Girls to Consider becoming Technicians in Engineering*, EITB

Howard G, Roberts P and Clift S, (1986) 'Attitudes Towards Women Scale (AWSB) Comparison of Women in Engineering and Traditional Occupations with male engineers', *British Journal of Social Psychology*, Vol. 25, pp 329-334

IPRA, (1991) *Future Skill Needs of the Construction Industries*, ED Report

IRS Employment Trends, (1989) *Recruitment Problems in the Engineering Industry*, IRS Employment Trends, No. 4440, 23 May, pp 5

Keenan T and Logue C, (1985) Interviewing Women, *Personnel Management* (IPM) Vol. 17, No. 4, April, pp 59

Kirk-Walker S, (1994) *Women in Architecture 1992: A synopsis of the main findings of the report*, Institute of Advanced Architectural Studies, September 1994

Latham M, (1993) *Trust and Money*: Interim report of the joint government/industry review of procurement and contractual arrangements in the United Kingdom construction industry, Construction Industry Council, December

Latham M, (1994) *Constructing the Team*, Final report of the joint government/industry review of procurement and contractual arrangements in the United Kingdom construction industry, HMSO, London

Latham Review Working Group 7, (1994) *Construction Industry Image: Proposal for a Twenty-First Century World Class Image for Construction*, Constructors Liaison Group, London

Lindley R M and Wilson R (eds), (1993) *Review of the Economy and Employment: Occupational Assessment*, Institute for Employment Research, University of Warwick

Lindley R (ed.), (1994) *Labour Market Structures and Prospects for Women*, Equal Opportunities Commission, Manchester

Lindley R and Wilson R (eds), (1994) *Review of the Economy and Employment: Occupational Assessment*, Institute for Employment Research, University of Warwick

Lorenz C, (1990) 'Kicking against the pricks', *Building Design*, 19 October, pp 24-25

Lorrimean J, (1987) 'Wise Careers for Women', *Training Officer*, Vol. 3, No. 10, October, pp 308-310

Lowe L and Byrne S, (1993) *His House or Our House?* A Report of the Women in Building Consultative Committee, The Chartered Institute of Building, December

Matthews A, (1994) *Gender and perceptions of job competence*, Research and Development Report 17, NCVQ

McLellan A, (1994) 'Latham calls for equal opportunities push', *New Builder*, 5/12, August, p 7

McRae S, (1990) *Keeping Women In: Strategies to Facilitate the Continuing Employment of Women in Higher Level Occupations*, Printer Publishers, Oxford

McRae S, (1991) 'Occupational Change over Childbirth: evidence from a national survey', *Sociology*, 25:4, pp 589-605

McRae S, Divine F and Lakey J, (1991) *Women into Science and Engineering*, Policy Studies Institute, London

McRae S and Daniel W W, (1991) *Maternity Rights in Britain: First Findings*, London, Policy Studies Institute

Meikle J, (1995) 'Image Problem', *Education Guardian*, 17 January

Millet E, (1994) 'Women pessimistic over prospects', *Contract Journal*, 17 November, p 11

Mills H, (1993) 'Legal slump hits women's prospects', *Independent*, October 26, pp 9

NACETT, (1994) *National Advisory Council for Education and Training Targets: Report on Progress*, NACETT, London

National Science Board, (1993) *Science and Engineering Indicators 1993*, United States National Science Foundation, Washington

New Builder (1994), 15/22 July

Newton P, (1987) 'Who Becomes an Engineer? Social Psychological Antecedents of a Non-Traditional Career Choice', Spencer A and Podmore D, (1987) *In a Mans World: Essays on Women in Male Dominated Professions*, Tavistock Publications, London

Opportunity 2000, (1994) *Third Year Report*, Opportunity 2000, London

Patel K, (1994) 'The women who surf on male turf', *Times Higher Education Supplement*, December 2, pp 18-19

PCAS, (1994) *Statistical Supplement to the PCAS Annual Report 1992-1993,* Universities and Colleges Admissions Service, Cheltenham

Peacock S, (1986) 'Enginering Training and Careers for Women in Britain: An Overview of the Initiatives Undertaken by the Engineering Industry Training Board', *European Journal of Engineering Education* Vol. 11, No. 3, pp 281-294

Pearson J, (1991) *Women in Construction*, The Chartered Institute of Building Direct Membership Examination thesis, May

Personnel Management, (1994) 'Women still face bottleneck on way to top jobs in personnel', August

Phillips D, (1987) 'Skills Shortages and the Pressures on the Building Industry', *Managing Construction Education*, Coombe Lodge Report, FESC, Vol. 20, No. 2, pp 79-87

Pickering C and J Woolard (1994), 'Women and men working effectively together', *Training Officer*, 30:8, October, pp 243-247

Pieniazek G, (1984) 'A Woman's Place', *Building*, 7 September, No. 7359 (36), pp 25-26

Pollert A and Rees T, (1992) *Equal Opportunity and Positive Action in Britain: Three case studies*, Warwick Papers in Industrial Relations, No. 42, IRRU, University of Warwick

PSI, (1994) Employers' Role in the Supply of Intermediate Skills, Policy Studies Institute, London

Platt Baroness, (1985) 'Women into Science and Engineering', *Equal Opportunities International*, Vol. 4, No. 1, 1985, pp 11-15

Rees T, (1994) *Women and the Labour Market*, Routledge, London

RIBA, (1983) *Women Architects*, RIBA, London

Rice R, (1991) 'Law Society reports rise in number of women solicitors', *Financial Times*, October 15, pp 11

Robinson G and McIlwee J S, (1989) 'Women in Engineering: A Promise Unfulfilled?', *Social Problems*, December, Vol. 36, No. 5, pp 455-472

Rubery J and Fagan C, (1994) 'Occupational Segregation: Plus can change...?', in R Lindley (ed.), *Labour Market Structures and Prospects for Women*, Equal Opportunities Commission, Manchester

Savage M, (1992) 'Women's expertise and men's authority: gendered organisation and the contemporary middle class', in Savage M and Witz A (eds), *Gender and Bureaucracy*, Blackwell, Oxford

Sheppard T, (1991) 'Can Women Make Good Engineering Managers?', *Women Engineer*, Vol. 14, No. 13, Spring 1991, pp 27-28, 31

Sims D, (1994) *GNVQs in Construction and the Built Environment: Final Report*, NFER/CITB

Sleep C A, (1994) *Female School Leavers into Construction*, A study for BSc Construction Management, Construction Management Unit Department of Surveying, University of Salford, March

Sly F, (1993) 'Women in the labour market', *Employment Gazette*, November, pp 483-502

Sly F, (1994) 'Mothers in the labour market', *Employment Gazette*, November, pp 403-413

Smithers A and Robinson P, (1994) *The Impact of Double Science in Our Schools*, Engineering Council, London

Sommerville, J, Kennedy P and Orr L, (1993) 'Women in the UK Construction Industry', *Construction Management and Economics*, 11, pp 285-291

Sommerville J and Stocks R K, (1993) *Reducing construction conflict: engineering the psycho productive environment*, proceedings of 9th Annual ARCOM Conference, September 14-16, Oxford University, pp 279-289

Srivastava A, (1992) *A Case Study of Widening Access to Construction Higher Education*, Proceedings of ARCOM Eighth Annual Conference, September 18-20, Douglas, Isle of Man, pp 230-239

Stone H, (1992) 'Senior Management — Are women fairly represented in the construction industry?', *Women in Construction*, papers presented at the conference held at the University of Northumbria, 10 September

Stoney S M, (1984) *Girls Entering Science and Technology: the Problems and Possibilities for Action as Viewed from the GE Perspective*, Paper Given to EOC, Manchester Polytechnic Girl Friendly Schooling Conference, September

Summers D, (1991) 'No room for new faces at the top', *Financial Times*, May 7

Summers D, (1992) 'Female directors report widespread inequality', *Financial Times*, March 19, pp 9

Syedain H, (1991) 'Women on their Metal', *Management Today*, February, pp 56-57, 59-60

Union of Construction Allied Trades and Technicians, (1989) *Blueprint for Equality: A UCATT Plan for Equality for Women in the Construction Industry*, UCATT, London

Unknown, (1993) 'Sexism on the A Gender', *RIBA Journal*, Vol. 100, No. 7, pp 11, July

Unknown, (1984) 'Jobs for Women: Public Sector Leads', *Architects Journal*, Vol. 179, No. 4, pp 32

Unknown, (1990) 'Building Design', *Woman's Realm*, October 19, pp 24-27

USR, (1994a) *First destinations of university graduates 1992-93*, Universities' Statistical Record, Cheltenham

USR, (1994b) *University Statistics 1993-94: Students and staff*, Universities' Statistical Record, Cheltenham

USR, (1987) *University Statistics 1985-86: First destinations of university graduates*, Universities' Statistical Record, Cheltenham

Van Den Berghe W, (1986) *Engineering Manpower: A Comparative Study on the Employment of Graduate Engineers in the Western World*, UNESCO, France

Watts J, (1995) 'Architects: last of the die-hard sexists', *Independent on Sunday*, 29 January, p 10

Whyte J, (1986) *Girls into Science and Technology*, London, Routledge

Wigfall V, (1980) 'Architecture', in Silverstone R and Ward A, *Careers of Professional Women*, Croom Helm, London

Wilkinson S, (1990) *Construction and the Recruitment of Female Labour*, Oxford Brookes University, School of Construction and Earth Sciences

Wilkinson S, (1992a) 'Looking to America: How to Improve the Role of Women in Civil Engineering', *The Woman Engineer*, Vol. 14, No. 16, Spring, pp 12-13

Wilkinson S, (1992b) *Career Paths and Childcare: Employers Attitudes towards Women in Construction, and Women in Construction*, papers presented at the conference held at the University of Northumbria, 10 September

Wilkinson S, (1993) *New Initiatives in Construction Management Education*, Proceedings 9th Annual ARCOM Conference, Oxford, September 14-16

Wilson R A, (1994) 'Sectoral and Occupation Change: Prospects for Women's Employment', in R Lindley (ed.), *Labour Market Structures and Prospects for Women*, Equal Opportunities Commission, Manchester

WISE, (1984) *WISE: Education and Training: Conference Report*, October

WISE Campaign, (1994) *WISE Vehicle Programme: An Engineering Council Project*, WISE, London

Witz A, (1992) *Professions and Patriarchy*, London, Routledge

Women in Construction Advisory Group, (1987) 'A Women's Place?' *National Builder*, Vol. 68, pp 342-343

Women in Construction Advisory Group, (1988) *Recruiting and Employing Women: A Guide for Construction Employers*

Women's Education in Building (n.d.), *Women's Education in Building*, London

Yeates J K, (1992) 'Women and Minorities in Engineering and Construction', *Cost Engineering* 34:6, pp 9-12

Young B, (1991) 'Insight: Educational Initiatives That Make Business Sense', *The Woman Engineer*, Summer, pp 13-14

Appendix 1: Methodology

The research comprised four main stages:

- a review of existing literature on women working in technology and science, with a particular focus on the engineering and building industries
- a postal survey of women with experience in the building industry
- group discussions with respondents to the postal survey, a group of careers advisers and building educationalists, and a group of employers
- telephone interviews with key industry figures.

A1.1 Literature review

The main aim of this stage of the research was two-fold:

- to provide a context for the rest of the study by summarising existing data and research on women in the construction industry and other male dominated activities
- to contribute to the discussion on recommendations for change in building via an analysis of how engineering, another male dominated industry, had approached the issue of women's under-representation.

The literature review drew on research from a range of sources including industry and academic journals, conference papers, published studies and unpublished research.

A1.2 Postal survey

The postal survey aimed to generate information on the characteristics and views of women with experience of the building industry. There were three distinct groups of women in the survey:

- Current members of the CIOB
- Former members of the Institute who had left within two years prior to August 1994 (this group was surveyed in order to gain a better understanding of the reasons women leave the industry)

Table A1.1 Response to the Women in Building survey

Outcome	All		CIOB members		Former members		Others	
	No.	%	No.	%	No.	%	No.	%
Participation	**468**	**63.8**	**282**	**76.8**	**97**	**49.7**	**89**	**52.0**
Non-participation	6	0.8	3	0.8	1	0.5	2	1.2
Inappropriate	15	2.1	1	0.3	6	3.1	8	4.7
No reply	244	33.3	81	22.1	91	46.7	72	42.1
Total mailed[1]	733	100.0	367	100.0	195	100.0	171	100.0

[1] A total of 44 Post Office Returns (PORs) have been excluded from the data on the number of questionnaires mailed. By group, there were five PORs among questionnaires sent to current members of CIOB, 27 among former members, and 12 in the 'Other' group. If PORs are included in the response rate, participation in the survey declines slightly to 60 per cent. The main groups affected are former members and the 'Other' group.

Source: IES Survey 1994

- Women in neither of the first two categories, who indicated they were willing to participate in the study by responding to an article in the building press or who were on building related courses, including the Building Industry Technical Training Scheme.

These three groups gave us a sample of 777 women. Questionnaires were sent out on 22nd September 1994 and, after two reminders, by the end of December, 489 responses had been received. Of these, 21 (three per cent) indicated that they did not wish to participate in the study or did not think it was appropriate for them to do so. Details of the response by group are shown in Table A1.1.

The response from current members of the CIOB was particularly high (over 75 per cent) which means that the results reported in the main body of the report can be said to represent the views of the majority of women in the Institute. A smaller proportion of former members and the 'Other' group responded (50 and 52 per cent respectively), which, given the nature of the samples, was only to be expected.

The questionnaire is reproduced as Appendix 2. It covered a number of topics, including:

- personal and educational characteristics
- reasons for entering the building industry
- views on building courses (where appropriate)
- views on the building industry
- reasons for leaving the industry (where appropriate)
- the effectiveness of a range of actions which could be taken to improve women's representation in building and their career progress within it.

The women surveyed were also encouraged to write about their reaction to the study and give further details of their experience in the industry and space on the questionnaire was left for this purpose.

A1.3 Group discussions

In total we conducted six group discussions with three distinct categories of participants. These included:

- four groups of women working in the industry (a total of 24 participants). In three of these cases participants were drawn from respondents to the postal survey, who had indicated their willingness to participate in a follow-up discussion on the postal questionnaire. In the fourth case, the participants were women who were attending a Women in Building Consultative Committee meeting.
- one group of building industry careers advisers and educationalists (two participants). These were identified and contacted by the project Steering Group, which organised the event. IES researchers facilitated the discussion.
- one group of employers (six participants). These were also identified and contacted by the project Steering Group who organised the event on behalf of the IES researchers running the discussion.

The discussions focused on two main topics:

- how best to encourage more women to think of a career in the building industry
- policies which would help women already in building professions to remain within the industry and to progress in their careers.

They were held in Ascot, London, and the North West (Warrington) and lasted between one and two and a half hours.

A1.4 Telephone interviews

We were interested to obtain information on the likely industry reaction to the research findings. We therefore interviewed seven key industry figures over the telephone for an average of about 20 minutes. These discussions focused on the findings of the study and the feasibility of introducing policies to address some of the issues raised.

Key industry figures were identified by the project Steering Committee. The CIOB then sent an initial letter asking individuals to participate in the research. This was followed up by IES staff, who arranged a convenient date and time for the interviews to take place.

Appendix 2: The Questionnaire

Improving Prospects for Women in Building

The Chartered Institute of Building is sponsoring an initiative to improve the prospects for women working in the building industry. A key aim of the study is to develop concrete recommendations on the most effective way of attracting women to and retaining them in the industry.

This an important opportunity for women in building to make their views known; we are anxious to ensure that the results reflect the concerns and experiences of as many women as possible. Equal Opportunities is one of the key issues raised by Sir Michael Latham in his recent report "Constructing the Team". An industry Task Force is being formed to take this forward, and the outcomes of this research will be passed on to that Task Force.

The study is being conducted by IMS, an independent, not-for-profit research institute based at Sussex University.

Understanding of the roles, motivations and views of women with experience (even if it is past experience) of the building industry is a central part of the project. We are therefore contacting professional women in the industry, students, present and past members of the CIOB.

I am writing to ask you to complete the attached questionnaire, which should take about 20 minutes to complete. Not all of the questions will be relevant to you as the survey is designed to take a variety of experiences into account. However, the path through the questionnaire is clearly indicated: please complete it as fully as possible and return it directly to us in the reply-paid envelope.

I can assure you that all information will be completely **confidential** to IMS and will not be released to other bodies under any circumstances, nor will it be used by us for any purpose other than this study.

By completing this questionnaire you can contribute to improving the working environment in the building industry. Please do not hesitate to call me should you have queries about this survey. My number is 0273 686751. Thank you for your help, I look forward to receiving your response in the near future.

Gill Court

Gill Court
Project Director

September 1994

Improving Prospects for Women in Building

Section A — Personal Characteristics

1. What was your age last birthday?........................ years

2. What is your ethnic origin? (*please tick one box*)
   - 01 ❑ White
   - 02 ❑ Black-Caribbean
   - 03 ❑ Indian
   - 04 ❑ Black-African
   - 05 ❑ Pakistani
   - 06 ❑ Black-Other
   - 07 ❑ Chinese
   - 08 ❑ Bangladeshi
   - 09 ❑ Other (*please describe*) ....

3. Do you have any dependent children? ❑ Yes ❑ No

   If 'yes', how many are: ❑ Under 5 ❑ 5-16 ❑ Over 16

4. Are you caring for an elderly relative or other adult? ❑ Yes ❑ No

5. What is your marital status? (*please tick one box*)
   - 1 ❑ Married/Living with partner
   - 2 ❑ Single
   - 3 ❑ Divorced/Separated/Widowed

6. Did either of your parents work in a professional/technical capacity? *(please tick one box)*

   ❑ Mother only 1 ❑ Father only 2 ❑ Both 3 ❑ Neither *(Go to Q7)* 4

   Was this in the construction industry? *(please tick one box)*

   ❑ Mother only 1 ❑ Father only 2 ❑ Both 3 ❑ Neither 4

Section B — Educational and Professional Qualifications

7. We are interested in your educational and professional qualifications. Please indicate by ticking the relevant boxes in **Column A** your current highest academic, vocational and professional qualifications. If you are still studying or expecting to obtain additional qualifications please indicate the qualification(s) you are expecting to obtain by ticking the relevant box(es) in **Column B.**

	Current Highest *(Tick one per section)* **A**	**Expected** *(Tick all that apply)* **B**
Academic Qualifications		
None	❑ 1	❑
GCSE/Standards/'O' Level/CSE	❑ 2	❑
'A' Levels/Highers/'AS' Levels	❑ 3	❑
First Degree	❑ 4	❑
Higher Degree (*eg* MSc, MA, MBA, PhD)	❑ 5	❑
Other (*please specify*)	❑ 6	❑

Vocational Qualifications

None	❑ 01	❑
First Certificate/First Diploma/Intermediate GNVQ	❑ 02	❑
N/SVQ Level 2	❑ 03	❑
NC/ND/Advanced GNVQ/OND/ONC	❑ 04	❑
First Line Supervisors	❑ 05	❑
N/SVQ Level 3	❑ 06	❑
HND/HNC	❑ 07	❑
SMETS	❑ 08	❑
N/SVQ Level 4 or Level 5	❑ 09	❑
Other (*please specify*)	❑ 10	❑

Professional Qualifications

None	❑ 1	❑
CIOB Associate Exam	❑ 2	❑
CIOB Member Exam Pt I	❑ 3	❑
CIOB Member Exam Pt II	❑ 4	❑
Other (*please specify*)	❑ 5	❑

8. Are you a member of any professional organisations? ❑ Yes ❑ No
If 'yes', which one(s)

....

....

9. Have you undertaken any women only courses in building? ❑ Yes ❑ No

10. If 'yes', did this course lead to a qualification? ❑ Yes ❑ No

Section C — Career Choices

11. We would like to know about your decision to enter the building industry, in particular the factors which positively influenced your decision and any negative influences, or discouragement, you experienced. Using the scale provided, please indicate how the following factors influenced your decision to pursue a career in building by circling the appropriate number next to each statement.

Strong positive influence	Positive influence	Not an influence/ not relevant	Negative influence	Strong negative influence
1	2	3	4	5

An employer	1	2	3	4	5
A family member or friend already in the industry (*eg* family business)	1	2	3	4	5
Family or friends generally	1	2	3	4	5
Careers advisers in school or college	1	2	3	4	5
Careers advisers from the Construction Careers Service/CITB	1	2	3	4	5
Teacher or tutor in school or college	1	2	3	4	5
Careers publications, including videos	1	2	3	4	5
Contact with Chartered Institute of Building members or staff	1	2	3	4	5
Visits to building sites from school or college	1	2	3	4	5
General media representation of the industry	1	2	3	4	5
Work experience or job shadowing in the building industry	1	2	3	4	5
Financial rewards (salary, car, healthcare, childcare)	1	2	3	4	5
Job security	1	2	3	4	5
Status	1	2	3	4	5
Interested in the subjects or type of work involved	1	2	3	4	5
Good opportunities for career development	1	2	3	4	5
It offered an opportunity to start my own business	1	2	3	4	5
Wanted to work in a job not traditionally done by women	1	2	3	4	5
Wanted to work in varied job locations/out of doors	1	2	3	4	5
It's something I always wanted to do	1	2	3	4	5
It seemed like a good thing to do at the time	1	2	3	4	5
Other *(please specify)*	1	2	3	4	5

....

Section D — Your Current Job and Career History

12. Are you currently working in the building industry? ❑ Yes ❑ No *(Go to Section E)*

If 'yes', how long have you been working in the industry? ....(years)

13. What is your primary activity or area of interest? *(please tick one box)*

01	❑ Building Control	13	❑ Health and Safety
02	❑ Building Surveying	14	❑ International/Europe
03	❑ Building Technology/Materials	15	❑ Maintenance/Refurbishment
04	❑ Commercial Management	16	❑ Managing Construction
05	❑ Computers and IT	17	❑ Personnel Management
06	❑ Contracts/Legal	18	❑ Programming/Planning
07	❑ Design	19	❑ Project Management
08	❑ Development	20	❑ Purchasing
09	❑ Education/Training	21	❑ Quality Management
10	❑ Environmental Issues	22	❑ Quantity Surveying
11	❑ Estimating	23	❑ Research
12	❑ Facilities Management	24	❑ Other *(please state)* ....

14. What type of employer do you work for? *(please tick one box)*

1 ❑ Consultant 2 ❑ Contractor 3 ❑ Public sector

4 ❑ Educational establishment 5 ❑ Self-employed 6 ❑ Other *(please specify)* ....

15. Do you mainly work: ❑ Off-site 1 ❑ On-site 2 ❑ An equal mix of the two 3

16. Is your job: ❑ Full-time ❑ Part-time

17. Are you based in the UK? ❑ Yes ❑ No

18. Is your husband (or partner) in paid employment?

❑ Not applicable *(Go to Q20)* ❑ No *(Go to Q20)* ❑ Yes

19. If 'yes', how would you describe the importance of their career compared to yours when household decisions are being considered?

1 ❑ Greater importance given to my career

2 ❑ Greater importance given to my husband/partner's career

3 ❑ Equal importance given to both our careers

20. Which of the following facilities or arrangements are available in your organisation (*please indicate all those available by ticking the relevant boxes in* **Column A**). In **Column B**, please indicate whether you would anticipate taking up the facilities listed if they were available to you. In **Column C**, please indicate whether you have already taken up any of the facilities listed.

	Available **A**	Would take up if available **B**	Have taken up **C**
Maternity leave (beyond statutory requirement)	❑	❑	❑
Career break schemes	❑	❑	❑
Job-sharing	❑	❑	❑
Part-time working	❑	❑	❑
Term time working	❑	❑	❑
Flexitime	❑	❑	❑
Emergency time off (*eg* for domestic reasons)	❑	❑	❑
Ability to work from home	❑	❑	❑
After school care/holiday play scheme	❑	❑	❑
Financial support for caring for dependents	❑	❑	❑
Creche	❑	❑	❑
In-house training/development (*eg* N/SVQs)	❑	❑	❑
Professional updating/retraining (*eg* after a career break)	❑	❑	❑
Financial support for external training	❑	❑	❑

21. We are interested in any career breaks you have taken while employed in the building industry. For this study, a career break is a break from work, however short, to have, or care for, a child or dependent adult. It includes maternity leave, shorter periods of maternity or adoption leave, and longer breaks for child rearing or other caring purposes.

a. Are you currently on a career break? ❑ Yes (*Go to Q22*) ❑ No

b. Have you ever taken a career break while employed in the building industry?
❑ Yes (*Go to Q23*) ❑ No (*Go to Q24*)

22. If you are currently on a career break:

How long do you expect it to last? ❑ Years ❑ Months

How old were you when it began? .... Years

Which of the following statements best describes your plans:

1 ❑ I plan to return to work for the same employer

2 ❑ I plan to return to work in the building industry but for a different employer

3 ❑ I plan to return to work but not in the building industry

4 ❑ I don't plan to return to work

(*please go to Q24*)

23. If you have had a career break:

How many career breaks have you taken? ....

How old were you when you began your **most recent** career break? .... years

After your **most recent** career break, did you return to work to the same employer?

❑ Yes ❑ No

While on your **most recent** career break were you kept up to date with relevant changes in your field?

❑ Yes ❑ No

If 'yes', was this paid for: ❑ By yourself ❑ By your employer

Section E — Building Studies

24. Which of the following statements best describes your situation? *(please tick one box)*

1 ❑ I'm engaged on a course of study related to building ***(Go to Q25)***

2 ❑ I'm **not** engaged on a course of study related to building but I'm working in the building industry ***(Go to Section G)***

3 ❑ I'm **not** engaged on a course of study related to building and **neither** am I working in the building industry ***(Go to Section F)***

25. Is your course: ❑ Full-time ❑ Part-time

26. How long is the total length of your course? .... years

27. Are you sponsored by an employer? ❑ Yes ❑ No

28. Does your course include a period of work experience? ❑ Yes ❑ No

If 'yes', how long a period of work experience is required? .... weeks

29. Have you completed your work experience? ❑ Yes ❑ No

30. We are interested in your views on your course. Please indicate the extent to which you agree or disagree with the following statements about your course by circling the appropriate number next to each statement.

Strongly agree	Agree	Neither agree nor disagree	Disagree	Strongly disagree
1	2	3	4	5

Good careers guidance is provided	1	2	3	4	5
The work is challenging and interesting	1	2	3	4	5
There is too much emphasis on maths and science subjects	1	2	3	4	5
The work is too demanding generally	1	2	3	4	5
The environment is supportive of women students	1	2	3	4	5
The tutors and lecturers take me seriously	1	2	3	4	5
Fellow students take me seriously	1	2	3	4	5
My choice of specialism was influenced by my tutors/lecturers	1	2	3	4	5

I would feel more comfortable if there were more women tutors and lecturers	1	2	3	4	5
I would feel more comfortable if there were more women students	1	2	3	4	5
I was well prepared academically for the course	1	2	3	4	5
The course environment is too aggressive/competitive	1	2	3	4	5
The course gives a good preparation for working in building	1	2	3	4	5
I would like more women-only classes/tutorials	1	2	3	4	5
The facilities on the course are good	1	2	3	4	5
The course provides a good induction to working **on-site**	1	2	3	4	5
Work experience is readily available	1	2	3	4	5

31. Are you planning to pursue, or continue with, a career in the building industry?

❑ Yes *(Go to Section G)*   ❑ No *(Go to Q 36)*   ❑ Don't know *(Go to Section G)*

Section F — Leaving the Building Industry

We are particularly interested in knowing about decisions to leave the building industry and would be grateful if you could answer the following questions.

32. Are you working in the construction industry, but not the building sector? ❑ Yes *(Go to Section G)*   ❑ No

33. If 'no', are you currently in paid employment?

❑ Full-time 1   ❑ Part-time 2   ❑ Not in paid employment 3 *(Go to Q36)*

34. If you are in paid employment, could you briefly describe the type of work you currently do?

....

35. In your current job, do you use the skills and knowledge acquired during your training and work in the building industry?

❑ Yes   ❑ No

36. What factors influenced your decision not to pursue or continue a career in the building industry? Please indicate the importance of the following factors by circling the appropriate number next to each statement.

Extremely important	Very important	Quite important	Not important	Not relevant
1	2	3	4	5

A. Wanted to spend more time with my family	1	2	3	4	5
B. Wanted to work in a less male dominated field	1	2	3	4	5
C. My promotion prospects were limited	1	2	3	4	5
D. Difficulties combining work and home life	1	2	3	4	5
E. Wasn't prepared for reality of working in building	1	2	3	4	5
F. Wanted a change of direction	1	2	3	4	5
G. Wanted to pursue personal interests	1	2	3	4	5
H. Lack of job satisfaction/status	1	2	3	4	5
I. Lack of training and development opportunities	1	2	3	4	5

J. No longer wanted to work such long hours	1	2	3	4	5
K. Wanted better pay and benefits	1	2	3	4	5
L. Sexual harassment at work	1	2	3	4	5
M. Wasn't able/willing to relocate frequently	1	2	3	4	5
N. Wanted to work in a more supportive environment	1	2	3	4	5
O. Wanted better facilities at work	1	2	3	4	5
P. Wanted more secure employment	1	2	3	4	5
Q. My progress to a senior position was blocked	1	2	3	4	5
R. Didn't like working on-site	1	2	3	4	5
S. Couldn't find work	1	2	3	4	5
T. Wanted to work is a less aggressive environment	1	2	3	4	5
V. Got tired of being the only woman around	1	2	3	4	5
W. Other *(please specify)*	1	2	3	4	5

37. Which were the 3 most important factors influencing your decision to leave the industry? Please indicate by putting the letter corresponding to the relevant factors (*eg* A for Factor A) in the boxes below.

❑ Most important ❑ 2nd most important ❑ 3rd most important

38. Do you plan to return to work within the building industry in the future? ❑ Yes ❑ No

Section G — Attracting and Retaining Women Builders

This section asks about your views on the building industry and the most effective actions which could be taken to improve the representation and progress of women within it.

39. The following statements represent a variety of both positive and negative views on the building industry and women's position within it. Please indicate the extent to which you agree or disagree with the following statements by circling the appropriate number next to each one.

Strongly agree	Agree	Neither agree nor disagree	Disagree	Strongly disagree
1	2	3	4	5

The work is challenging and interesting	1	2	3	4	5
The pay for professionals in building is good	1	2	3	4	5
Women get a lot of hostility from their peers	1	2	3	4	5
Women get a lot of hostility from more senior people	1	2	3	4	5
Women get a lot of hostility from less senior people	1	2	3	4	5
There are too few women in building	1	2	3	4	5
The work is dirty and dangerous	1	2	3	4	5
It's harder for women to get promotion than men	1	2	3	4	5
There is a positive attitude to equal opportunities issues in the industry	1	2	3	4	5
The status of professionals in building is good	1	2	3	4	5
Working on-site is enjoyable	1	2	3	4	5
You have to be mobile to succeed in building	1	2	3	4	5

Building is a good industry for women who enjoy taking on non-traditional roles	1	2	3	4	5
Women can bring a less confrontational style to the industry	1	2	3	4	5
It's difficult working with people from such varied social backgrounds	1	2	3	4	5
The work is varied and demands a range of skills/abilities	1	2	3	4	5
Building employers are actively promoting equal opportunities	1	2	3	4	5
It's difficult to combine work and family life if you work in building	1	2	3	4	5

40. Please rank on a scale of 1-5 how effective you think the following actions would be in encouraging women **to enter** the building industry by circling the appropriate number next to each statement.

Extremely effective	Very effective	Quite effective	Not effective	Would **discourage** entry by women
1	2	3	4	5

A. Campaign to change attitudes of careers advisers	1	2	3	4	5
B. Make information on the range of careers available	1	2	3	4	5
C. Encourage girls to do maths and science	1	2	3	4	5
D. Counter view of building as unfeminine career	1	2	3	4	5
E. Give girls more building work experience opportunities	1	2	3	4	5
F. Introduce an insight programme, such as site visits, for girls	1	2	3	4	5
G. Introduce an insight programme on building for teachers/careers advisers	1	2	3	4	5
H. Offer more job security	1	2	3	4	5
I. Offer more settled job locations	1	2	3	4	5
J. Provide better information to parents	1	2	3	4	5
K. Provide better information to teachers	1	2	3	4	5
L. Highlight the role of women already in building	1	2	3	4	5
M. Offer women only courses	1	2	3	4	5
N. Get employers to encourage applications from women (*eg* in their adverts)	1	2	3	4	5
O. Counter the dirty/dangerous image of building	1	2	3	4	5
P. Highlight the good career prospects available	1	2	3	4	5
Q. Get the media to show a more positive view	1	2	3	4	5
R. Get building into the National Curriculum	1	2	3	4	5
S. Ensure careers advice given prior to age 14	1	2	3	4	5
T. Other *(please specify)*	1	2	3	4	5

41. Please indicate which you think would be the 3 most effective actions to encourage women to enter the industry by putting the letter corresponding to the relevant actions (*eg* A for Action A) in the boxes below.

☐ Most effective ☐ 2nd most effective ☐ 3rd most effective

42. Please rank on a scale of 1-5 how effective you think the following actions would be in encouraging women **to remain** in the industry by circling the appropriate number next to each statement.

Extremely effective	Very effective	Quite effective	Not effective	Would **discourage** retention
1	2	3	4	5

A. Involve women in developing equal opportunities policies	1	2	3	4	5
B. Offer women only training/development	1	2	3	4	5
C. Develop mentoring schemes within organisations	1	2	3	4	5
D. Provide support to develop networks for women	1	2	3	4	5
E. Ensure equal opportunities principles in selection, promotion, advertising, training etc are introduced	1	2	3	4	5
F. Monitor recruitment, retention, progress of women	1	2	3	4	5
G. Offer industry awards to successful women	1	2	3	4	5
H. Get more women trainers	1	2	3	4	5
I. Assertiveness training courses for women	1	2	3	4	5
J. Equal opportunities awareness training courses for men and women	1	2	3	4	5
K. Ensure objective assessment procedures	1	2	3	4	5
L. End informal recruitment/promotion practices	1	2	3	4	5
M. Raise the profile of successful women	1	2	3	4	5
N. Introduce/improve flexible working/childcare	1	2	3	4	5
O. Introduce/improve career break schemes	1	2	3	4	5
P. Just let women get on with their careers	1	2	3	4	5
Q. Offer support to young women entering building	1	2	3	4	5
R. Public commitment by the industry to equal opportunities (*eg* by joining Opportunity 2000)	1	2	3	4	5
S. Commitment to equal opportunities from senior managers	1	2	3	4	5
T. Induction courses which include equal opportunities awareness for all new recruits	1	2	3	4	5
U. Other *(please specify)*	1	2	3	4	5

43. Please indicate which you think would be the 3 most effective actions to encourage women to remain in the industry by entering the letter corresponding to the relevant actions (*eg* A for Action A) in the boxes below.

❑ Most effective ❑ 2nd most effective ❑ 3rd most effective

Section H — Further Contact

44. We will be holding a series of discussion groups with women in the building industry to explore further some of the issues raised in this questionnaire. Would you be willing to participate in such a discussion group?

❑ Yes ❑ No

45. If 'yes', please could you give us your name and a telephone number at which we could contact you.

Name: ....

Telephone number: .... ❑ Home ❑ Business

Section I — Additional Comments

Please use the space below for any other comments you wish to make about your views on women in the building industry:

Thank you for taking the time to complete this questionnaire. Please return it in the reply paid envelope to: Monica Haynes, IMS, Mantell Building, University of Sussex, Falmer, Brighton, BN1 9RF

Appendix 3: Additional Material

Table 4.9A Availability of flexible working arrangements and training related facilities by sector

	Available		Have taken up	
	Public sector employer	Private sector employer	Public sector employer	Private sector employer
Maternity leave (beyond statutory requirements)	83.6	31.0	11.0	1.9
Career break schemes	17.8	5.6	1.4	0.9
Job share	56.2	6.0	2.7	0.5
Part-time working	52.1	13.9	9.6	2.3
Term-time working	13.7	1.9	2.7	0.9
Flexitime	63.0	10.2	39.7	4.2
Emergency time off	60.3	55.1	17.8	14.4
Ability to work from home	12.3	15.7	8.2	6.5
After school care/holiday play scheme	12.3	2.8	1.4	0.9
Financial support for caring activities	1.4	0.9	1.4	0.9
Crèche	24.7	0.5	8.2	0.5
In house training/development (eg N/SVQs)	53.4	42.1	37.0	23.6
Professional updating or retraining	23.3	10.2	11.0	3.2
Financial support for external training	61.6	43.1	42.5	26.9
No cases	73	216	73	216

Source: IES Survey 1994

Note: A total of 289 responses to this question were analysed by sector of activity (public or private). For all women 299 responses were analysed. The discrepancy between these figures is accounted for by 10 women who did not provide information on whether they were in the public or private sector.

Figure 5.3A The importance of factors in women's decision to leave building: women aged under 25 (ranked by order of importance)

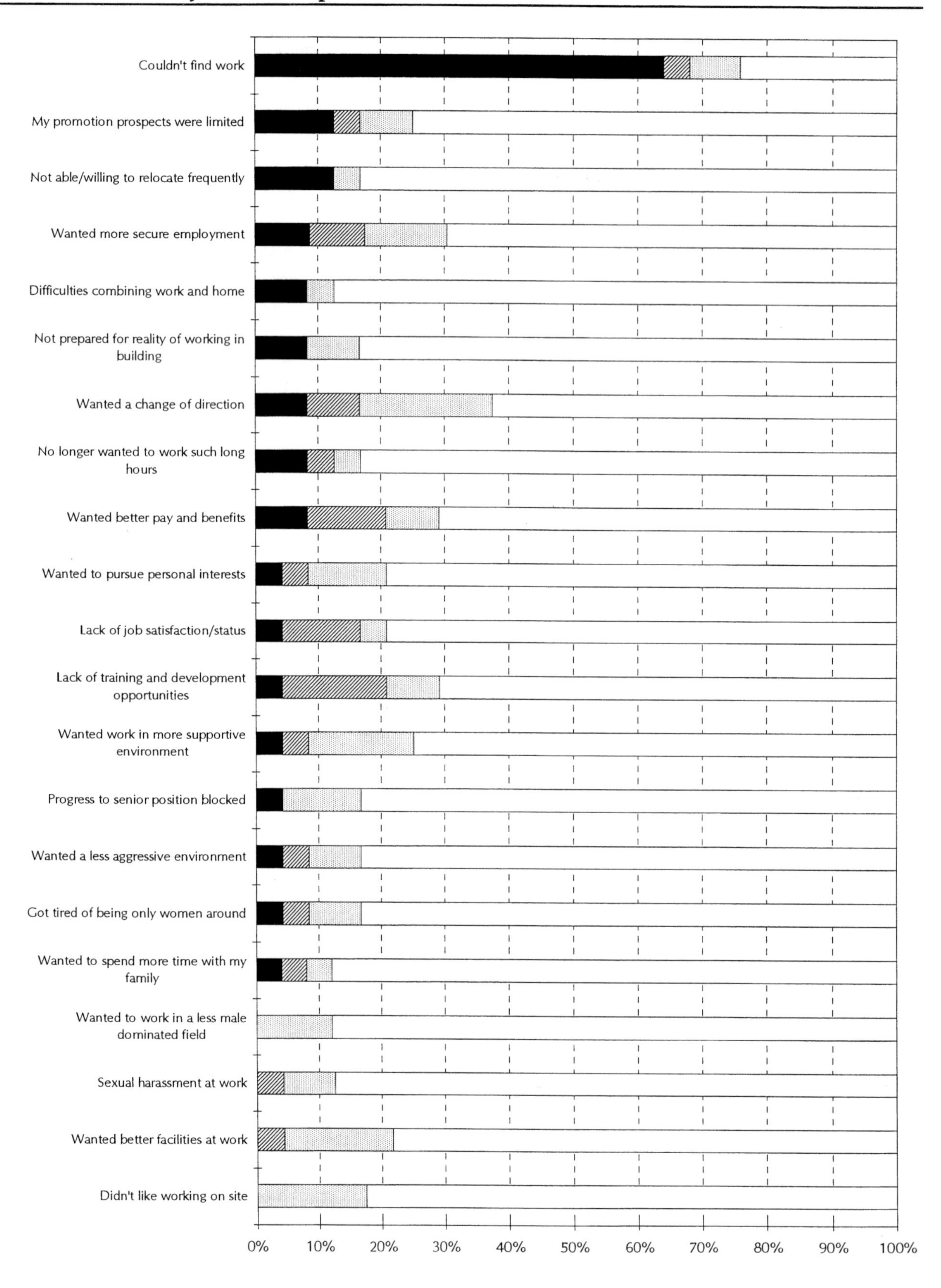

Source: IES Survey (see Appendix2 Question 36 for full text of statements)

Figure 5.3B The importance of factors in women's decision to leave building: women with children (ranked by order or importance)

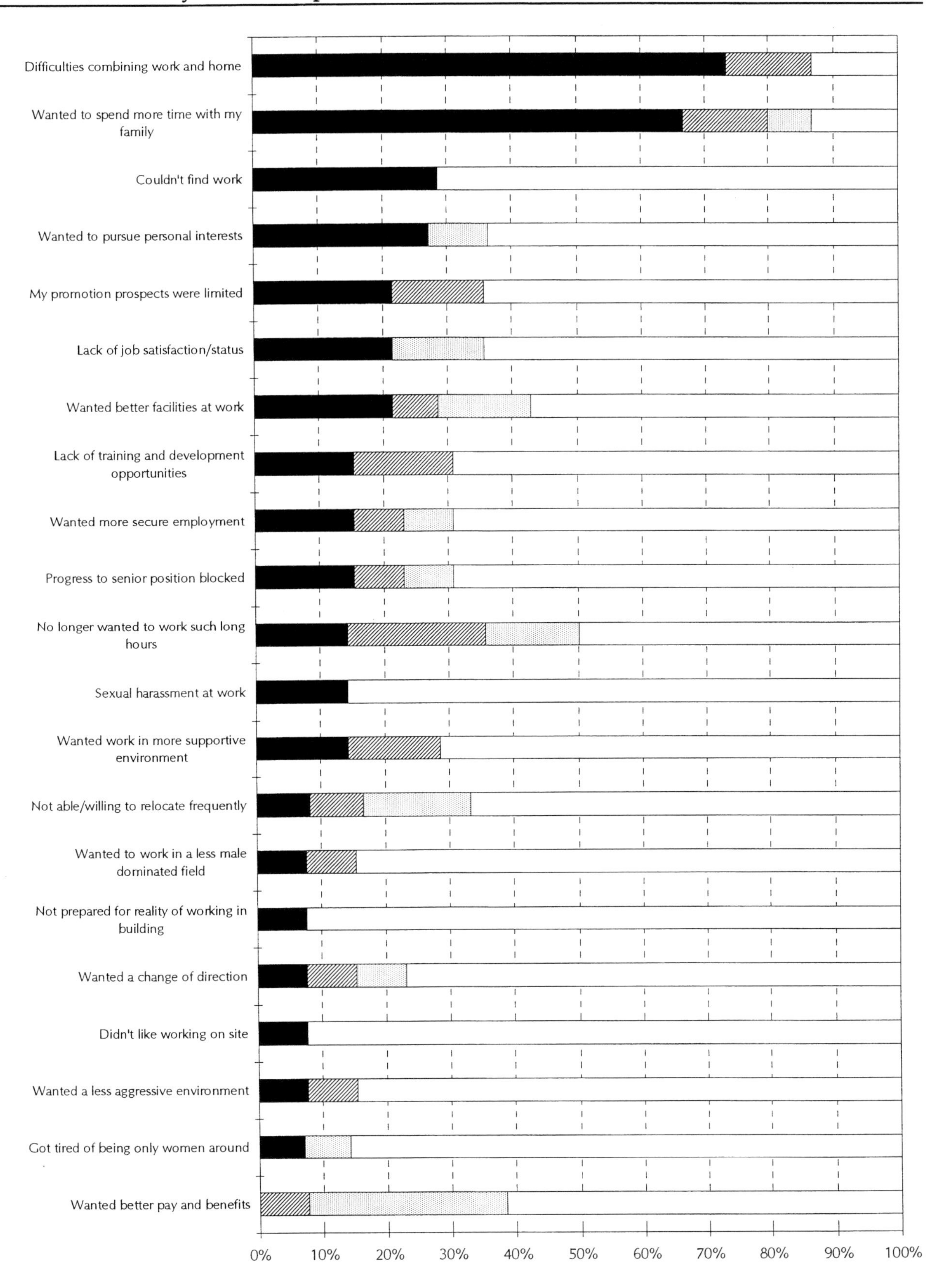

Source: IES Survey (see Appendix2 Question 36 for full text of statements)

Figure 6.1A Students' views on their courses: including neither agree nor disagree

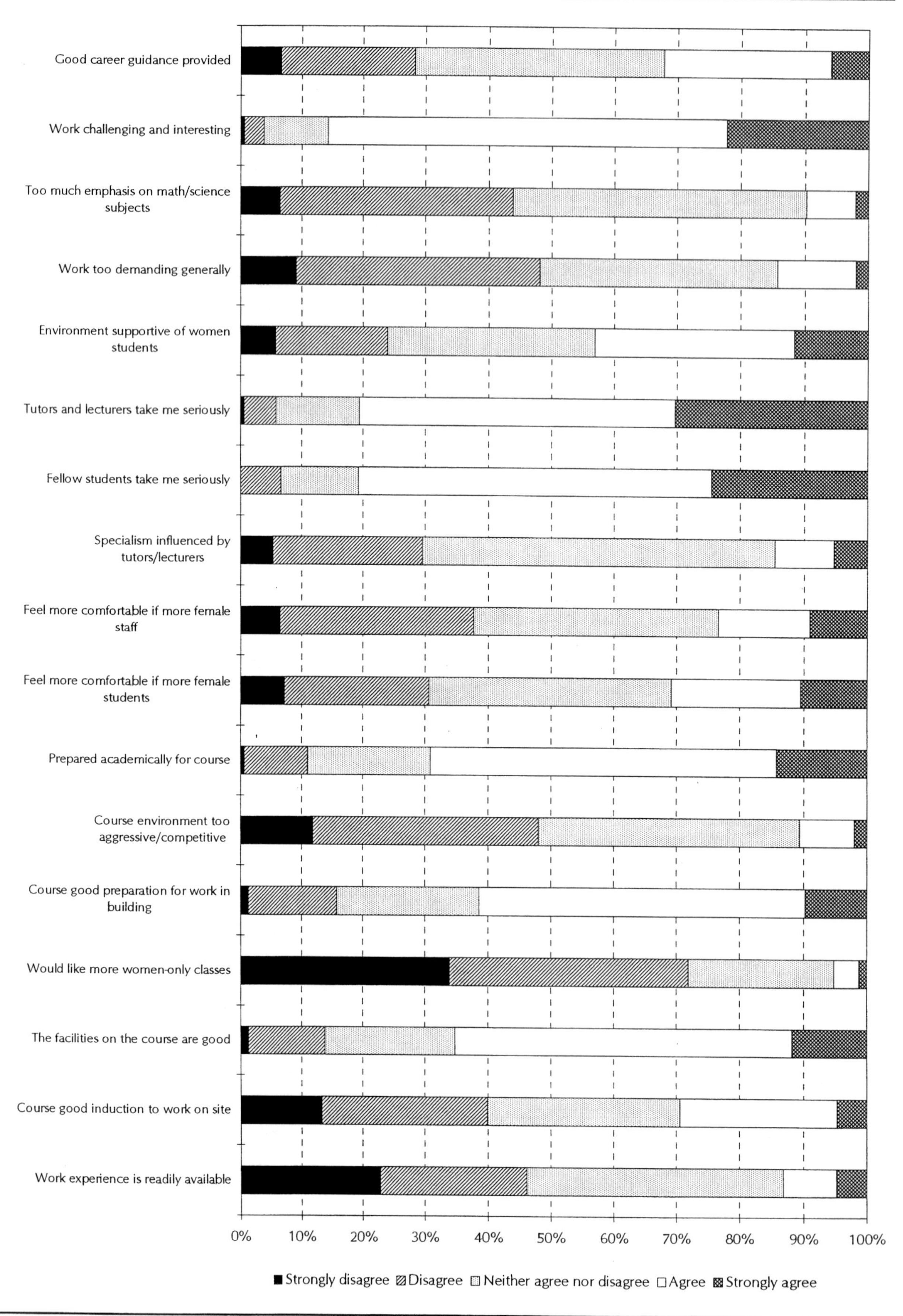

Source: IES Survey (see Appendix2 Question 30 for full text of statements)

Figure 7.1A Women's views on the building industry: all respondents (including neither agree nor disagree)

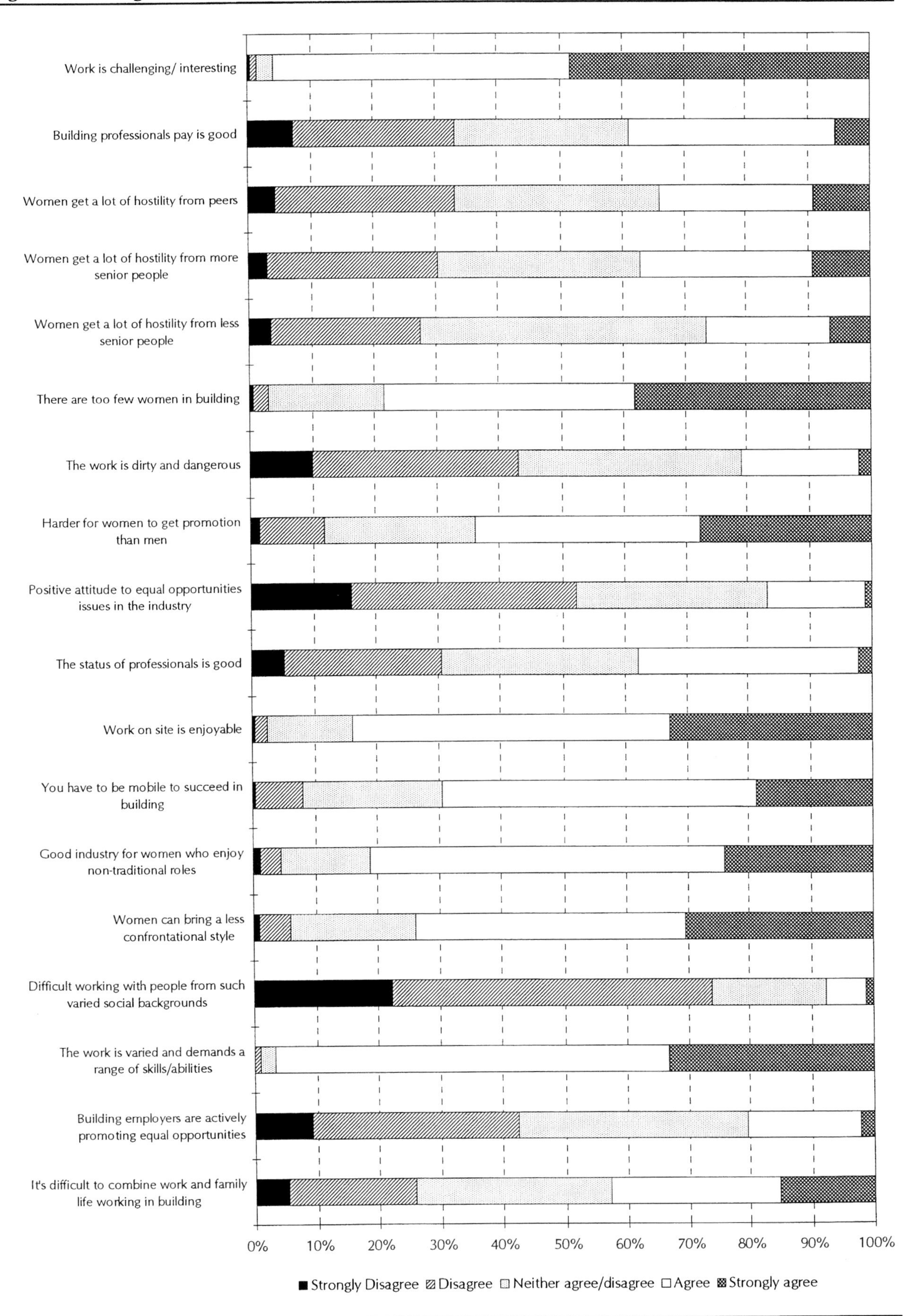

Source: IES Survey (see Appendix2 Question 39 for full text of statements)

Figure 7.1B Women's views on the building industry: CIOB members

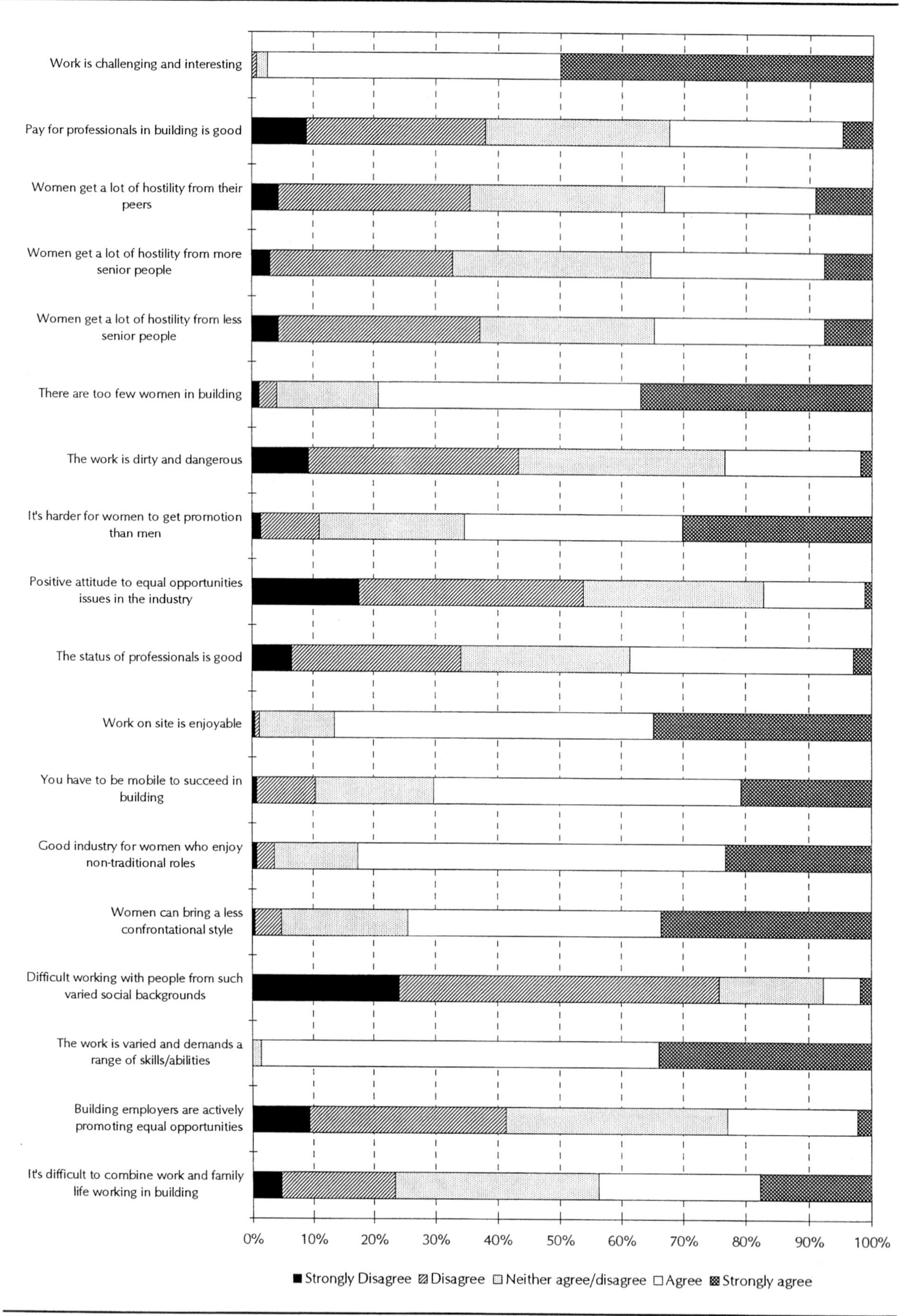

Source: IES Survey (see Appendix2 Question 39 for full text of statements)

Figure 7.1C Women's views on the building industry: former members

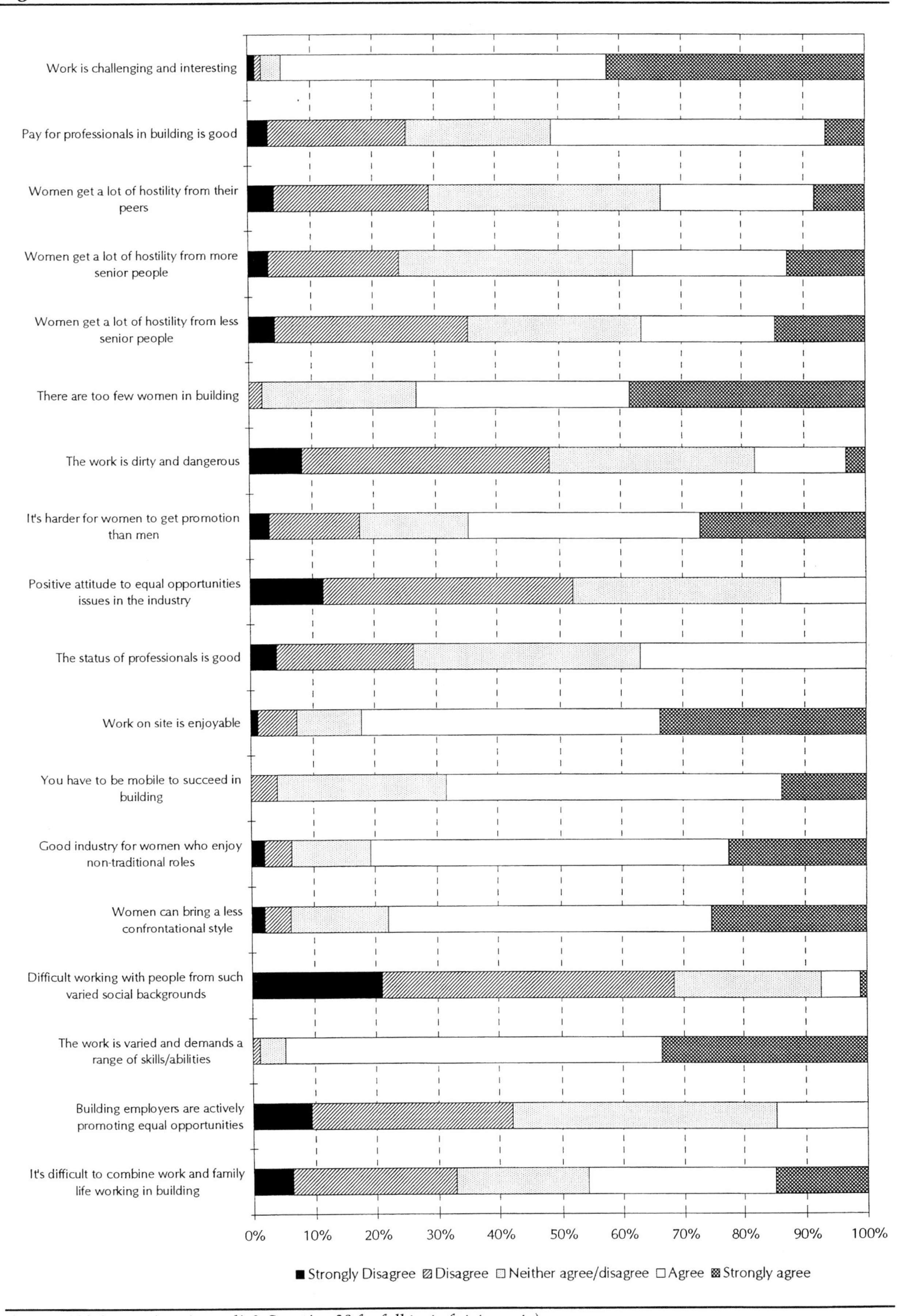

Source: IES Survey (see Appendix2 Question 39 for full text of statements)

Figure 7.1D Women's views on the building industry: 'other' group

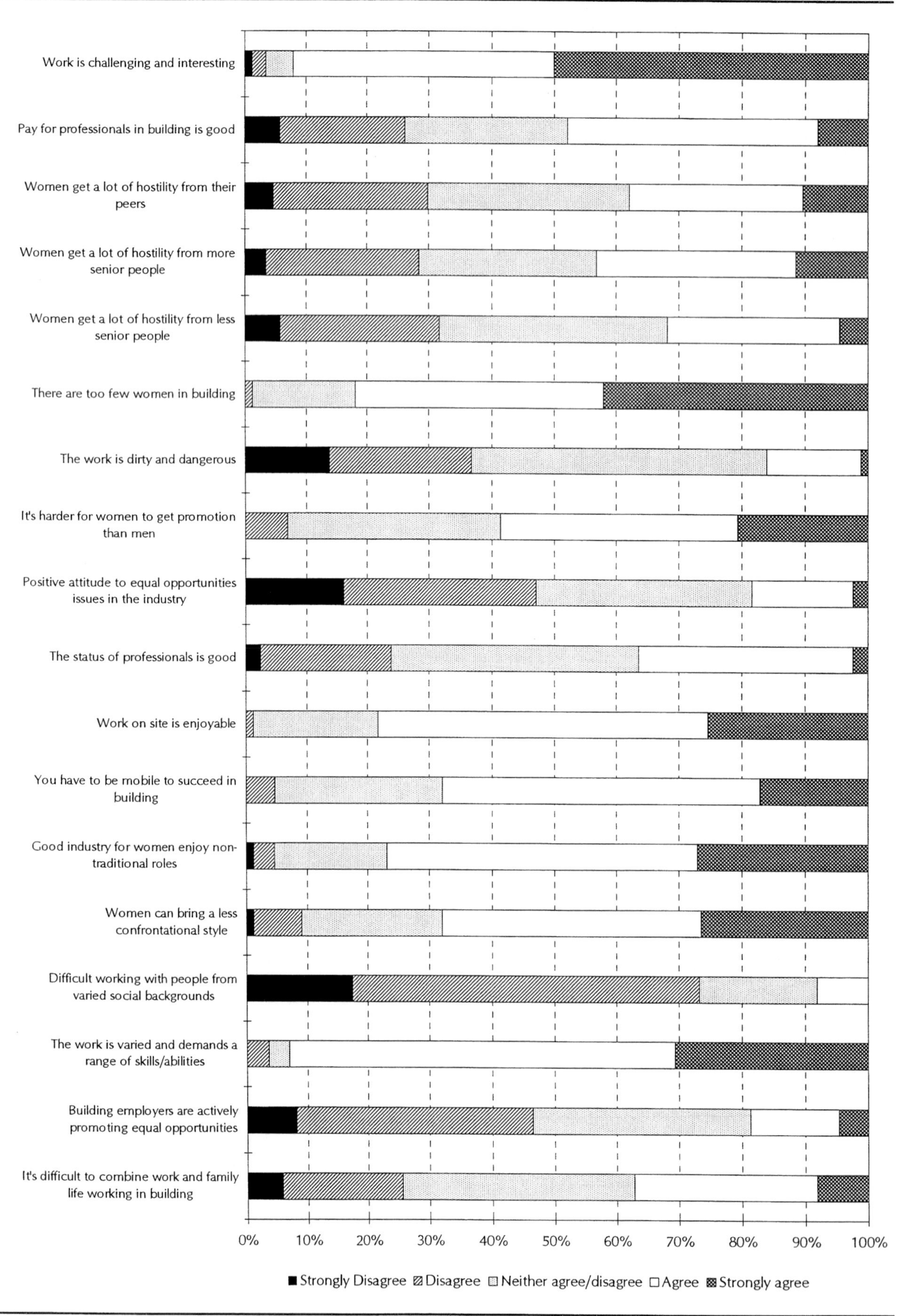

Source: IES Survey (see Appendix2 Question 39 for full text of statements)

Figure 7.2A Effectiveness of actions to encourage women to enter building professions: CIOB members

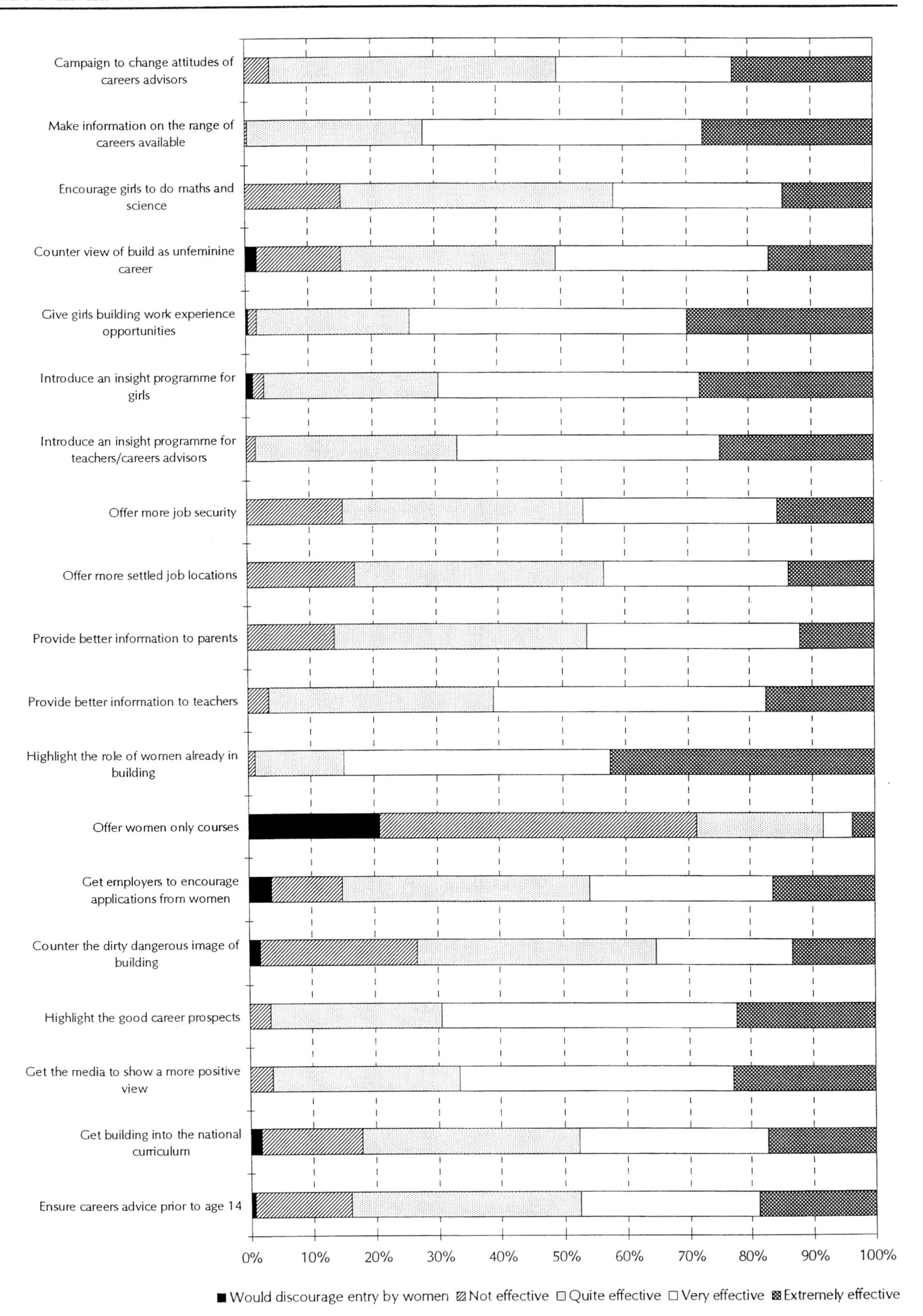

Source: IES Survey (see Appendix2 Question 40 for full text of statements)

Figure 7.2B Effectiveness of actions to encourage women to enter building professions: former members

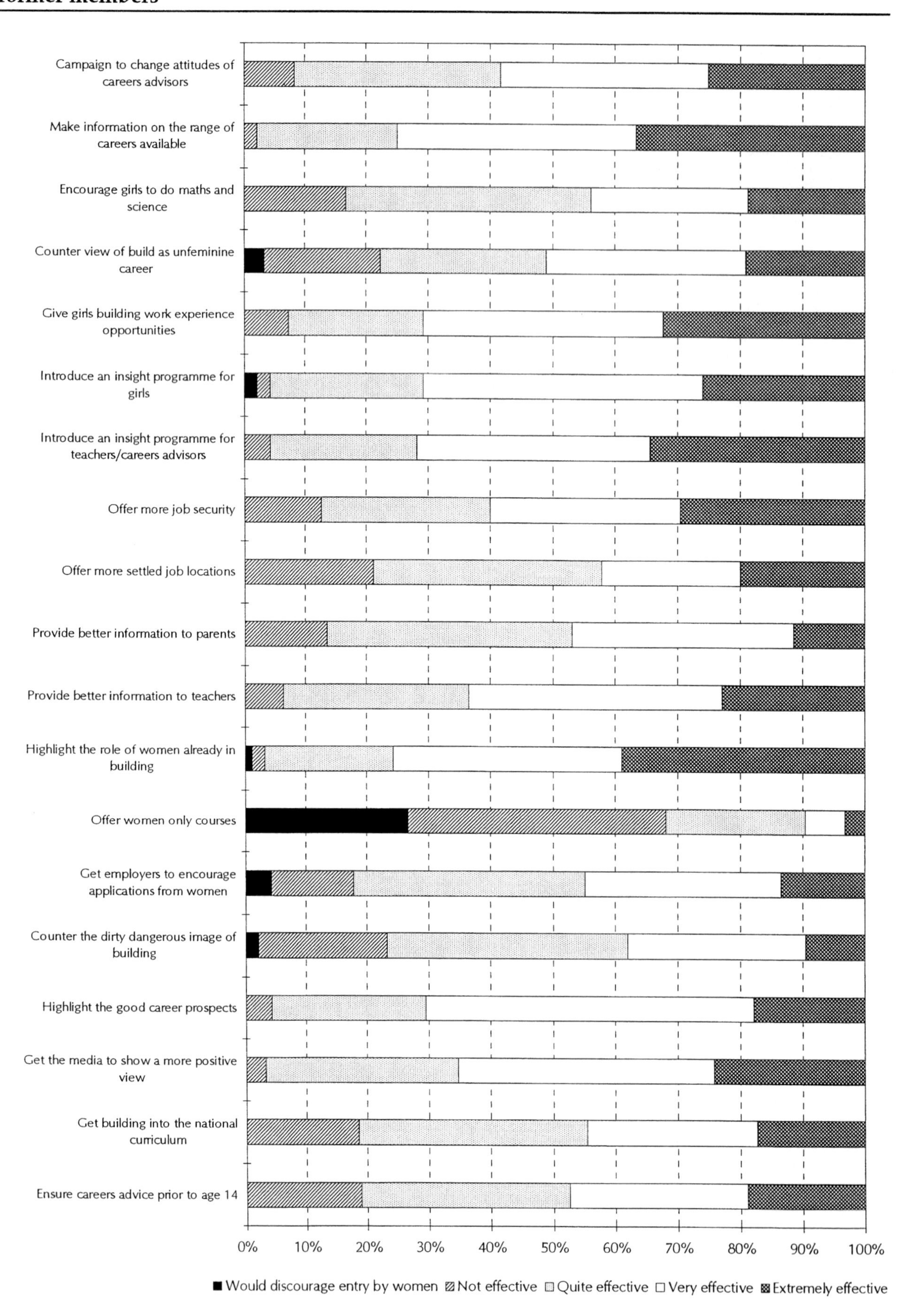

Source: IES Survey (see Appendix2 Question 40 for full text of statements)

Figure 7.2C Effectiveness of actions to encourage women to enter building professions: 'other' group

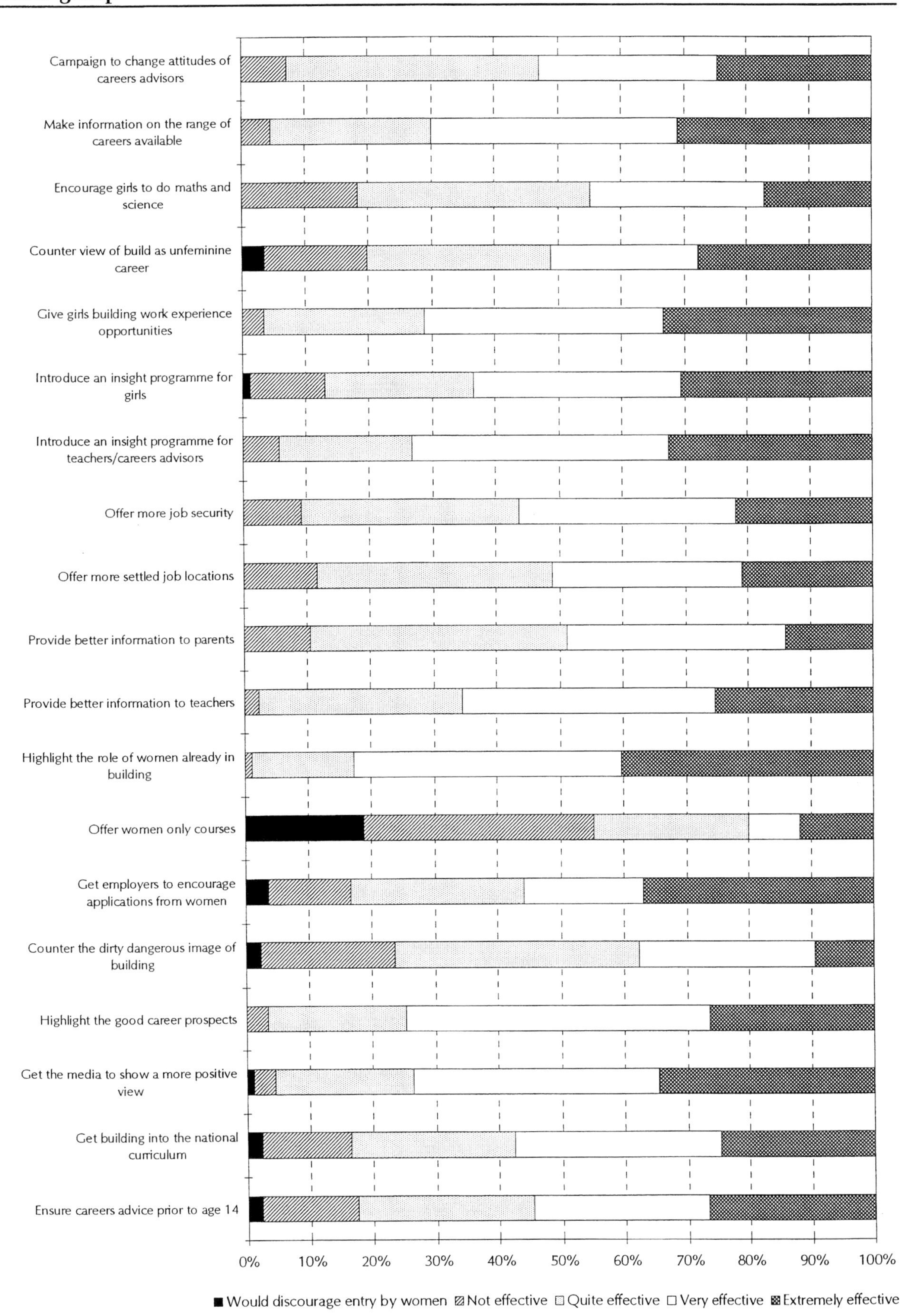

Source: IES Survey (see Appendix2 Question 40 for full text of statements)

Figure 7.3A The effectiveness of actions to encourage women to remain in building: CIOB members

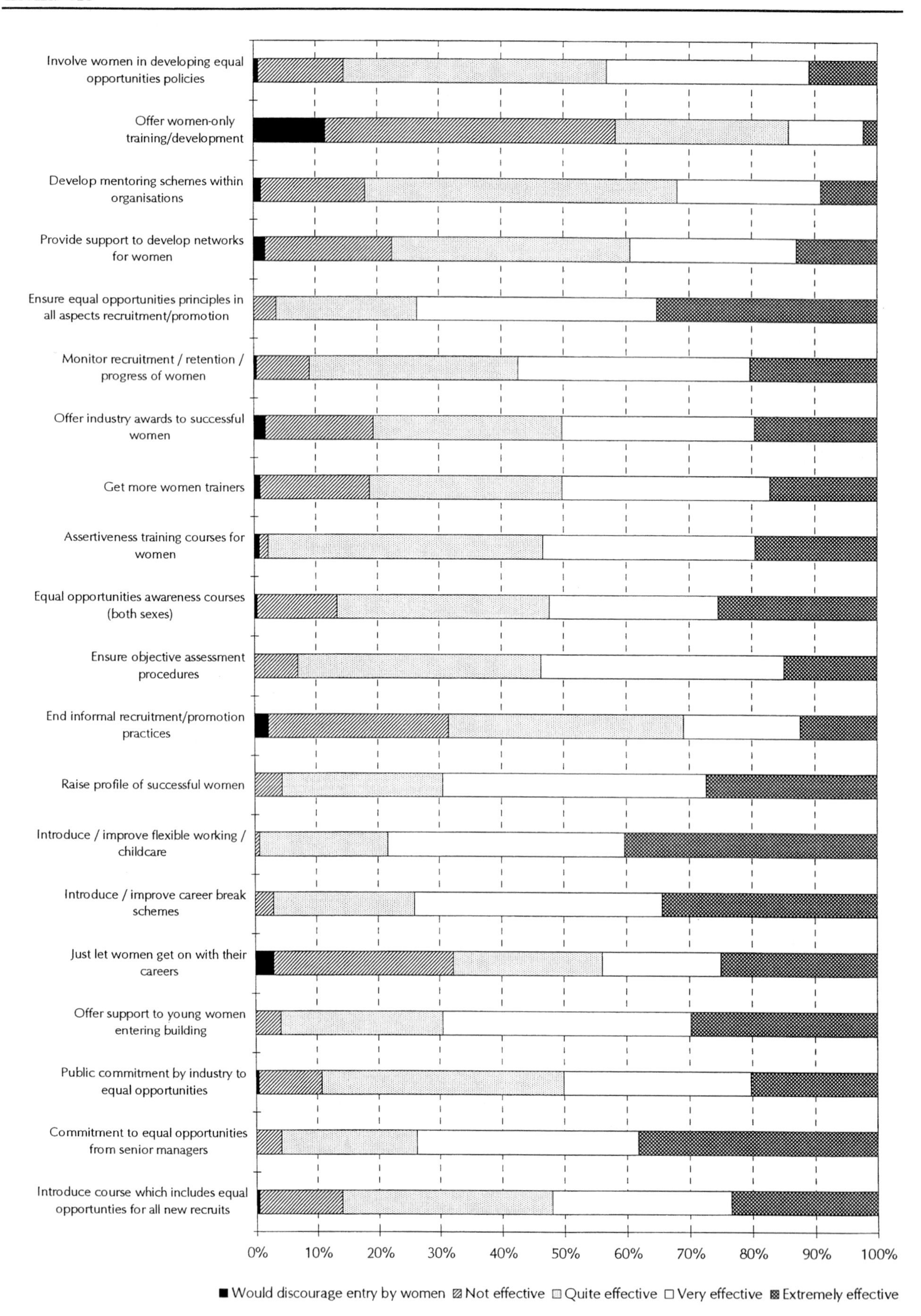

Source: IES Survey (see Appendix2 Question 42 for full text of statements)

Figure 7.3B The effectiveness of actions to encourage women to remain in building: former members

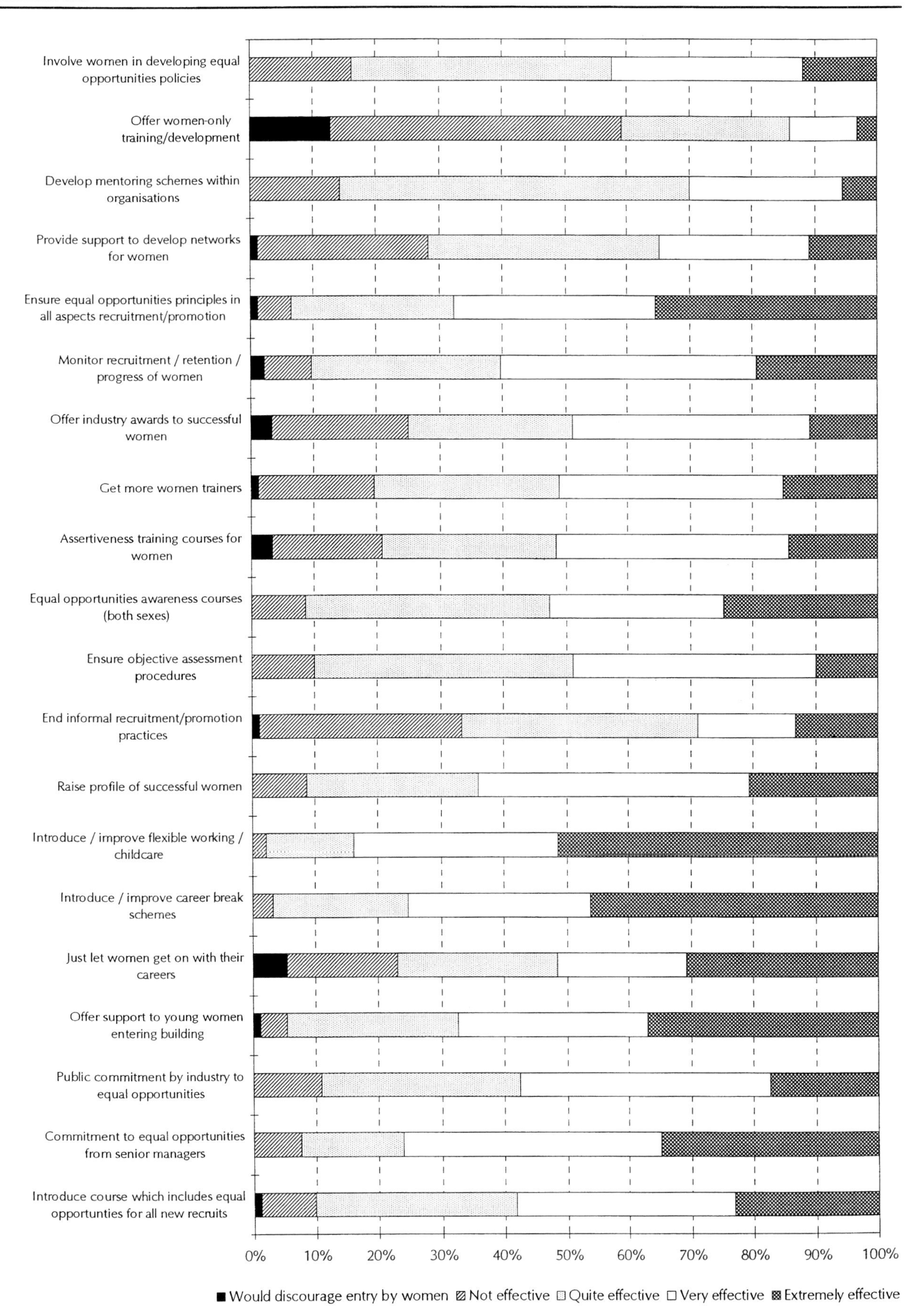

Source: IES Survey (see Appendix2 Question 42 for full text of statements)

Figure 7.3C The effectiveness of actions to encourage women to remain in building: 'other' group

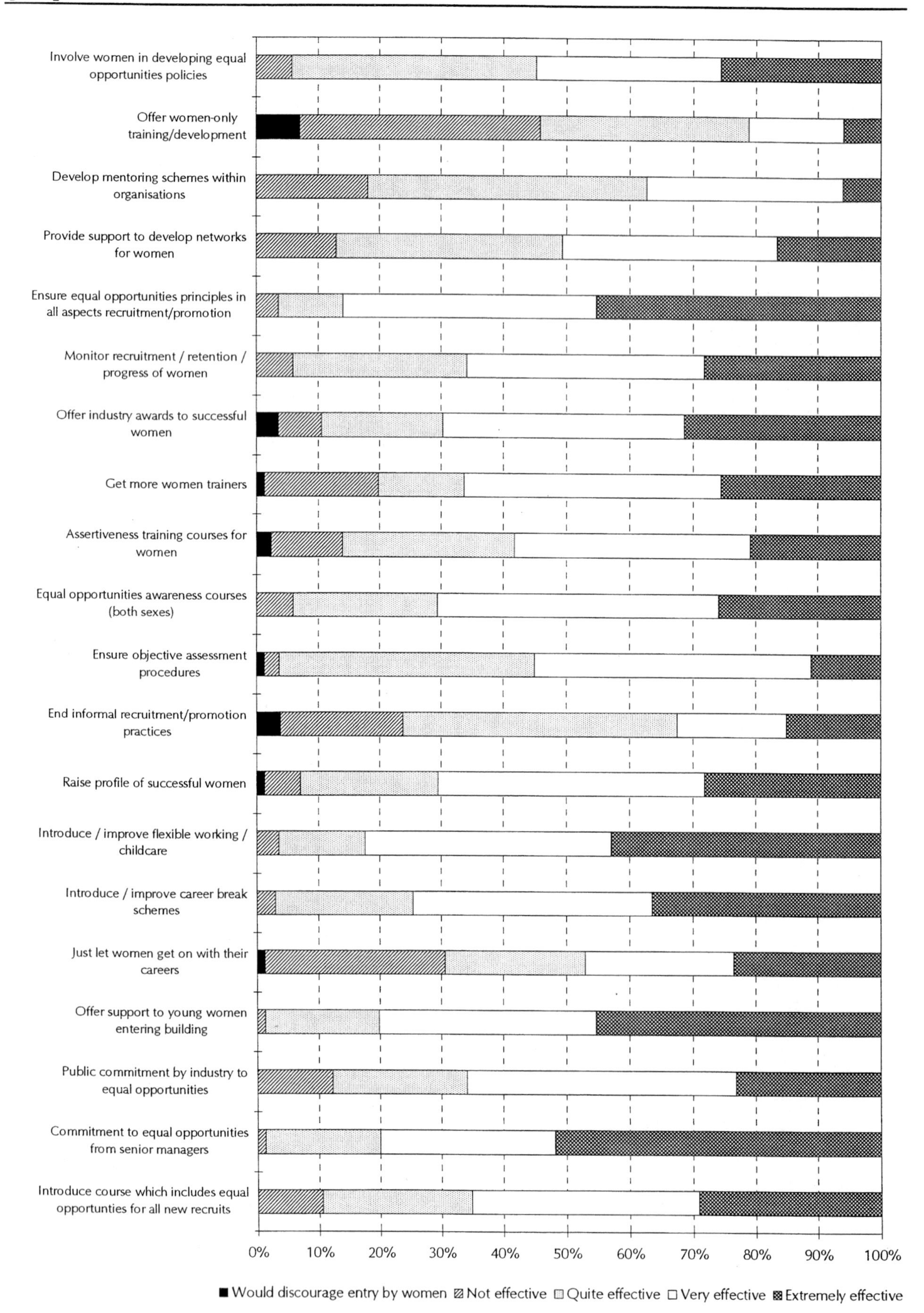

Source: IES Survey (see Appendix2 Question 42 for full text of statements)